W9-CHU-395

Unless a Grain of Wheat

Stephen Boehrer (signature)

A NOVEL

Stephen L. Boehrer

WIND-BORNE PUBLICATIONS

Copyright © 1997 by Stephen L. Boehrer

Cover Design by Judy Bridges and Sheila Studer

Library of Congress Catalog Card Number: 97-91102

ISBN: 0-9660607-0-9

Published by: Wind-borne Publications
               P.O. Box 733
               Hales Corners, WI 53130

First printing: October 1997
Printed in U.S.A.

TO RITA WITH LOVE

ACKNOWLEDGEMENTS

To my wife, **Rita**, voracious reader, for her insightful critiques and constant encouragement.

To **Judy Bridges**, Director of Redbird Studios in Milwaukee, for the craft. This book would not exist without her competent guidance and patient insistence on rewrite, rewrite, rewrite.

To the following for their critiques and support: Jane Eschweiler, SDS; John Heagle; James Patrick Shannon; Jo McReynolds-Blochowiak; Terry and Frank Ryan; Carol and Jack Letellier; Rita and James McDonald; Alice Kaiser and Pat Sheridan, FSPA; Rose Marie Ward and James Biechler; Betty and Bill Richard; Mary Ann and Jack Melloh; Audrey and George Hinger; Gene Bleidorn and Mary Agnes Blonien; Marjorie Pagel; and the writers at Redbird Studios.

To my nephews, James Kobs and the other Stephen L. Boehrer, for their technical support.

CHAPTER ONE

Bishop Frederick Patrick Sweeney looked out from the car's rear seat at the clapboard rectory, Mueller's rectory, a faded white, two-story box at the edge of the village. The driver opened the rear door and the bishop, briefcase in hand, stepped into the first few drops of what he sensed to be an all-day rain. He walked to the door and tried the knob. The door opened. He turned and waved to the driver. The car moved off.

Inside, the bishop paused at the door, breathing the unfamiliar odor of another man's home. A feeling of unease gripped him for a moment, though he knew the house was empty. The housekeeper had fled to stay with family.

Doors to the left and right led into equal-sized rooms, offices. He walked through the door on the right and was greeted by a 10x12 glossy of himself. In its black plastic frame, the portrait shared its wall with no one. He recalled sending a print to every parish in the diocese and wondered now, had it been pastoral, or vulgar? A table and chairs had been shoved into a corner, apparently, from the scrubbed look of the floor, to allow for cleaning. The opposite office was empty except for a confessional screen and chairs.

He moved farther into the house and entered the living room, a dining ell to his left. He winced momentarily at the dull plain appearance. Father Charles Mueller's passions lay elsewhere, certainly not in home furnishings.

The living room extended to the back of the house where a brown leather recliner guarded a bay window. He walked to the window and looked out at a mowed rectangle that surrounded a large garden. Bare spots in the garden indicated a harvest well underway. The bishop, tall like Mueller, sat in

1

the recliner and felt its comfort. He picked through a batch of books and magazines on a stand next to the chair. Surprised that Mueller would own Thomas A' Kempis' *The Imitation of Christ*, he flipped through its pages. Penciled judgments dotted the margins: "Good, Nah, Ugh, B.S." How like Mueller to challenge any authority, even the saints....

"I am the bishop of this diocese, Father Mueller. You may have your doctorate in theology, but I am the legitimate magisterium in this diocese. I'll have you remember that!" The bishop shifted slightly in the chair, feeling embarassment now at the recall of his imperial behavior.

The bishop had set it up himself, a program to talk about pastoral problems. He planned to take it on the road to reach all the priests in the diocese. He had invited Mueller to be a part. At the program's debut he and Mueller had stood in front of thirty priests.

Jack Casey, the pastor of St. Jude's, had said he wasn't hearing contraception in the confessional anymore. Should we be preaching on it? Mueller had told him to let a sleeping dog lie, that the Holy Spirit had talked directly to the people on that question, and that the *sensus fidelium*, the common sense of the people, had decided the issue.

"You don't say that until Rome says it first," the bishop had replied. Had he been offended at the thought that God might deviate from a divinely constituted chain of authority, or by the implied competition to his own authority?

Mueller had joked back that the Vatican had yet to admit that the earth revolves around the sun. "Galileo's been dead for three hundred fifty years. Priests need answers today."

The bishop had ended Mueller's role in the program right there.

Bishop Sweeney scanned the living room. The walls were barren except for a crucifix, a framed print of the Madonna of the Streets, and a pencil sketch of a gaunt bearded man, dressed in a loincloth, walking under a desert sun. He went to examine the sketch, sensing a familiarity. The title, *Jesus After Forty-Day Fast*, appeared at the bottom along with the artist's signature, *Eileen Fogarty*. He recalled the three Fogarty pieces that adorned the chancery walls when he first arrived in the diocese. He had them removed to a storage room after Eileen retired. He was told they now decorated the walls of the chancery ladies' room, put there, obviously, by staff secretaries from their affection for Fogarty.

The dining room table was piled high with papers, magazines and books. Where did the man eat? He walked into the kitchen and turned back immediately. Stairs led from the dining room to the second floor. He took them, and listened to the squeaking complaints of century-old lumber. A single wooden-floored hall divided the upstairs. He opened the first door. It was the housekeeper's apartment. He remained at the doorway and surveyed an open closet. Angela Pannetto would be returning for her clothes. Farther down the hall an open door revealed a guest bedroom.

He entered the door on the opposite side, Mueller's bedroom. A bed, dresser and a chair. A single bare nail stuck out of a wall. Had it held a painting, a photograph? A framed family photograph sat on the dresser. Ordination day? There was Mueller in cassock standing between his parents, a head taller than his father. A brother and a sister flanked the parents. The bishop picked up the photo to study it. Behind the picture lay a pair of dice....

"It has come to my attention, Father Mueller, that you gamble. I would add that if this knowledge has worked its way to me, it can only follow that you are gaining a reputation as a gambler." The bishop remembered the stress he had put on the word "reputation".

"I play a little poker and dice with some friends once a month," Mueller had responded, unconcerned, from the other side of the bishop's desk.

"Do you think a reputation for gambling fits the image a priest should present to his flock?" He had been insistent.

"Well, bishop, I know it doesn't make any difference to my card-playing friends. And I don't think of my parishioners as birds or sheep. I think they'll judge me on my merits."

"I think it's your job to be better than that, above reproach, like Caesar's wife."

It came out reluctantly. "Bishop, I've heard even bishops go to the track in mufti now and then."

Bishop Sweeney had been at the track in Arlington with friends. He had not gambled. But the appearance? Meetings with Mueller always seemed to end in a draw.

The bishop left the bedroom and walked through a bathroom that connected on the other side to a large room at the rear of the house, Mueller's study. An old oak desk, scored with age, faced him. Behind the desk, windows gave a view down onto the lawn and garden. A brown overstuffed chair sat to the left near a hall door, overseen by a tall lamp whose fraying cream shade had begun to brown. The right wall was lined with bookcases. Voids, some a foot wide, interspaced the books. Books missing? He walked behind the desk and sat, feeling the hardness of the cushionless chair.

This house waits, he thought, as if Mueller had merely gone to the store for groceries. Sitting here one would expect at any moment to hear the sound of Mueller's noisy old car in the drive. But Mueller would not be back. That much was certain.

The bishop felt he was at the center, in the control room of a man he never understood. Here, hopefully, were some answers to the questions that brought him here. Every time I met Mueller, the bishop recalled, he raised questions that I avoided. Why? Why didn't I look for answers? Obviously, I didn't value the questions. How many times was I face to face with this man who could prick the thick hide of a comfortable conscience? And yet I remained smug. If I'm honest about it, that's why I didn't like the man.

He had come unattended, to be alone, to search for understanding without the disturbance of other voices, and without their maliciously pious conjectures.

A dial phone squatted on the desk to the left of a large annotated calendar that served as a blotter. Stuck in a corner of the blotter was a piece of stationery with the handwritten salutation, "Dear Maggie." Maggie? A breviary rested on the front edge of the desk.

He opened the top right-hand drawer. Office supplies. The drawer below was filled with files. One folder was labeled "Fishing". All the others were labeled "Sermons", each with a specific year. The top drawer on the left held a variety of items: A sheaf of envelopes bound with a rubber band. He riffled through them. Each was stamped with a return address and postage. Each was sealed. All were addressed to Sister Margaret McDonough. Was she the "Maggie" of the salutation? If Mueller hadn't intended to mail them, why were they still here? The bishop thought for a moment, nodded his decision, and dropped the sheaf into his briefcase. He reached for the solitary piece of stationery and placed that, too, in the briefcase.

Bishop Sweeney sorted through penknives, sunglasses, a ring of keys, two padlocks, an address book, and a thin green ledger. He retrieved the ledger. "Game Record" was printed in ink on the label. It was a record of Mueller's gambling back to his years as a young associate. It was a casual record, entries scratched in pencil or pen at hand, regardless of color. Date, a plus or minus number, and a balance. The bishop followed the numbers down the several pages. The most money won on any one occasion was $3.75, the most lost, $3.10. The win/loss balance after a dozen or more years was a plus $28.40.

The bishop replaced the ledger and swiveled the oak chair to the window. From this height the hodgepodge of bare spots in the garden displayed a harvesting that was clearly methodical. He turned back to the desk.

The lower left file drawer was also filled. The front folder was titled, "Letters to Rome". It was empty. How many letters had there been? He knew of only one. On the rest of the folders the title, "Lecture", was followed by the name of a specific course of study. He closed the drawer.

His eyes fell on the breviary. He reached for it and flipped through the worn pages. A single sheet of paper, cut to page size, slipped from inside the cover to the desktop. It had been carefully hand printed in ink:

Credo: Ordination Day

I believe in God and in Jesus, His Son.

I believe in the Church.

I believe that I am called to obedience.

Today I made a solemn vow of obedience.

6

Penciled at the bottom was the word "over". He turned the page and read in scribbled pencil:

Modification: 10th Anniversary

Obedience to any church authority whose commands sacrifice people to the god of the institution is wrong and cowardly. God, give me the courage to break my vow rather than deny this new understanding which you have raised up in me.

Bishop Sweeney read both sides a second time, then swiveled again to the window. He did not see the rain. And vivid memories held him unaware of moisture welling in his own eyes.

CHAPTER TWO

From his youth Charles Mueller knew the importance of walking fences.

On this cool sun-bright September afternoon he walked the north fence line of his brother's farm. It was the home farm. He had spent his youth working this land, tending its cattle and walking these fences.

The farm lay on rich alluvial soil between the great river and the bluff. Sandstone cliffs made foreheads for the bluff's tree-bearded face. The cliffs shone yellow gold in the afternoon sun.

He walked the narrow strip of tall, heavy grass that separated the fence from a freshly plowed field. His direction followed the slow rise of the land toward the bluff. Halfway along the fence row he stooped, picked up a broken piece of sod and hurled it at a fence post. He tried to pray. Is it necessary, Lord, that you teach me what hell is by making me walk through it? Is this a test? I know what hell is. Hell is walking a fence line and finding the man I was a dozen years ago on one side of the fence and the man I am today on the other. Hell is watching your heart pasture on one side of the fence while your head grazes on the other. Which strayed first, my head or my heart? And how do I prevent a permanent divorce? Help me get them back in the same pasture. I want to stay!

A bevy of quail exploded from the grass in front of him and checked his thoughts. He stopped. His hazel eyes followed their flight to the tree line ahead.

His step was shorter as the grade increased toward the field's corner where a tiny patch of white suddenly caught his attention. A grub worm, fallen from a furrow's ridge into its valley, squirmed in quarter moons to right itself. Another movement led Charles's focus to the edge of the grass. There, a toad made a short wary hop toward the grub, then another hop, and another. It stopped a scant two inches from the grub. The toad cocked its head first to the right, paused, and then to the left. Suddenly, its tongue blurred. The grub disappeared.

Charles reached down, picked up the toad and carried it gently. He worked his long legs and narrow hips between the two upper fence wires at the corner of the field but snagged his shirt on a barb of the upper wire as he pulled his broad shoulders through. He freed the snag with his free hand and walked to a ledge of sandstone at the tree line. The ledge ran twenty feet across and rose four feet above the grass at its front. He climbed it easily on a grassy slope at its side. He walked onto the ledge and sat down, legs dangling over the edge. He set the toad down to his right.

The toad took two hops, to the edge. Then it squatted onto the sun-warmed rock, its body away from Charles, but its head cocked to the side and eyes alert.

Charles looked out over the land. His brother's home and farm buildings, the family homestead, rose from the soil to the left on the opposite side of the farm. His parents' new home lay to the right, on a lot carved out of a pasture. In the distance the great river ribboned its way south. He ran his fingers through a shock of auburn hair and looked at the toad.

The toad squatted, still.

Charles let his thoughts settle on the toad. You are a predator, Mr. Toad. We have predators where I work.

The toad hunched down on the rock, its eyelids closing and opening in a rhythm that looked like sleepy contentment.

Charles watched the bright sunlight glimmer a reddish purple on the toad's back and warts. Could it be that red and purple are the colors of predators?

The toad launched itself into a series of short hops and moved several feet from Charles before settling down again. It seemed torn between the rock table's warmth and the safety of the high grass.

Charles leaned back on his two arms. The predators in my organization devour their victims in other ways than you, Mr. Toad. Were you angry when you ate your neighbor a few moments ago? It didn't seem so to me. Charles's eyes narrowed. There is anger in my guts, Mr. Toad, and I am not a predator. It has taken me years to name that anger. A dozen years ago I was proud to wear the roman collar. Now?

The toad took several more hops toward the ledge's grassy side. It stopped about six feet from Charles.

I feel like I am being digested just as our friend, the grub, is being digested in your belly, Mr. Toad. Charles leaned out toward the toad. He judged that the toad's enjoyment of the sun gave way to its need for safety, as a series of hops carried the toad into the grass and out of his sight.

Charles sat quietly, absorbing the sun. Shadows flitted across the ledge. He looked up to see a pair of eagles soaring above the cliff. Eagles had thrilled him since he was a boy. He had more than once walked around the cliff to the summit and found their nest in a tall oak. Was it still in the same tree? They fed in the river and in the sloughs that bordered the farm. He could see clearly their white heads and tails. He watched them lift and glide in long lazy circles. And they gave a momentary lift to his spirit. That's what trust is like, he thought, giving yourself to the currents of air. Maybe what I need now is a daily supplement of trust. Monsignor "Frosty"

Oberkirche, Charles's boyhood pastor, had said that "trust is keeping your belly open to the sword." That was Frosty's kind of metaphor. I'd rather soar on eagle's wings, Charles thought.

Charles stood up and walked into the woods. A short while later he emerged at a lane that led to his parents' home. He stopped under a tree and jumped to get a hand grip on a limb. He pulled his weight until his chin was even with the limb, then felt the release of tension when he dropped to the ground. "When I got angry as a boy, Lord," he prayed aloud, "I didn't want my anger to go away quickly. I wanted those who caused my anger to feel it. I feared it wouldn't last long enough, and it didn't last. Something always came along to turn me from my anger and tease me into laughter."

He left the tree's shade for the sun and warmth of the lane. He walked toward his parents' home. Face to the sun he continued his prayer. "Help me hold my anger, Lord. Let me use it to fuel my engine. Let me hold it for those who turn your house into a sanctuary for predators of any kind."

Charles saw his father come around the corner of the garage a quarter mile ahead. He saw Lady, his dad's English setter, romp around the older man as they walked toward Charles. When the dog spied Charles it came charging. Charles knelt on the grassy land and the dog braked into him. He rubbed strong fingers behind Lady's ears as she tried to tongue-wash his face.

"You sure do get along with that dog, Son." Karl Mueller's suntanned face broke in a smile. "She don't take to just anybody."

Charles stood and took in the kind face of his father. "It's all those dog biscuits I bribed her with as a puppy, Dad. And they were your dog biscuits."

"Did you have a good hike?" Karl asked as they started walking to the house.

"I did. It's good for me to walk up to those cliffs now and then to get a reminder of my size. Afterwards I don't feel it's my job to fix everything that needs fixing. And I always learn a lesson from the eagles."

Karl leaned down and plucked a yellow blade of grass. His fingers toyed with the stem. "Your mother is worried about you, Son." He fixed the grass stem between his teeth.

"Mom worried? About me?" Charles understood the convention. His mother would say, "Your father is worried about you." They were both worried.

"You know how women are. They got this sixth sense." The grass was back in his fingers. "She doesn't think you look yourself. I know it sounds crazy, me asking a man who's going on forty, but is something troubling you, Son?"

"Nothing serious, Dad." Charles knew the answer was insufficient. "You know how jobs are. You get so wrapped up in it that you lose perspective."

Karl blew the grass stalk from his mouth, leaned down and picked another. "I got this feeling, Charles, there's something more to it." He paused and let silence turn the comment into a question.

If you haven't walked in my moccasins, Charles thought, how do I tell you what it's like. He stooped and picked a stalk of grass for himself. They walked in silence. Then, "Being a priest is a lot like working in the barn, Dad. Wonderful things happen there. But, there's also the barn smell."

"Aw, that smell ain't so bad. You get used to it." Karl wrinkled his nose.

"I know. But I'd feel I was selling my soul if I got used to the smell I'm talking about."

"Can you get rid of it?" Karl asked.

"I think it would take someone the size of those cliffs, or someone who can soar like the eagles," Charles answered. "But maybe I can help."

"Just don't think you have to do it all." Karl suddenly laughed. "Maybe it's a good thing neither of us has the best nose in the world." He reached a hand up onto his son's shoulder as they continued down the lane.

* * * *

Eva Mueller fussed. "Is the meat too done? Charles, you should take more mashed potatoes. What about you, Karl? Don't you like the salad?":

"Now you stop fishing for compliments, Eva. Everything is just fine," Karl teased.

Charles added his own praise. "You're still the best cook in the county, Mom."

Eva tasted each dish herself to make sure it met her standards. They were nearly finished eating when she put down her fork and looked at her son. "Do you ever think about getting married, Charles?"

Surprised, Charles looked at his mother and digested the question. He took time to give a reply. He thought about Maggie. "I'm sure everyone thinks about it, Mom. Why? You're thinking marriage is what's troubling me?"

Eva shrugged. "Perhaps."

"No, mother, if that were my problem, I wouldn't have a problem. I'd leave and get married. Now you stop worrying about me. I'm fine."

"Well," Eva persisted, "Dad and I just want you to know we want you to be happy."

As he drove the river road back to Mill Valley that night, his mother's words clung to his thoughts. For him to leave the priesthood, should he choose, was not a mortal sin against his mother, as it was for so many who left. Eva put first things first, her love for him. "Thanks, Lord, for those parents you gave me," he prayed. "And for their sake, help me keep this anger off my face. Help my face reveal the love in yours. And help me stay!"

CHAPTER THREE

A lifetime of cattle care and field work showed on the couple. Their chapped red hands and dark tanned faces told of endless hours in sun and wind. They entered the chancery office and hesitated in the large foyer. Neither wore a coat, though there was a chill in the spring air. A cotton print of pastel flowers covered the woman's broad expanse. The tall man wore an aged brown suit, a size too snug at the shoulders and sleeves. The knot of his tie was a hand-turned original, not unlike the knots he used for tethering cattle. A faint barn smell drifted from his clothes.

Josephine McGregor, Bishop Sweeney's secretary, saw them immediately, smiled, and beckoned them across the expanse of marble that separated the door from the elaborate mahogany enclosure that was her secretarial station. "Are you Mr. and Mrs. Krakowski?"

"Yah, we're Krakowskis. We got an appointment with the bishop," the woman responded. "We called ahead."

"I will tell the bishop that you are here." Josephine's smile failed to put them at ease.

"We come from St. Patrick's parish in Posnan; you know, over there by John's Point," the woman added, needlessly.

Josephine excused herself, left the enclosure, and walked into the bishop's office. "Your ten o'clock appointment is here, Bishop; Mr. and Mrs. Stanislaus Krakowski from St. Patrick's parish in Posnan."

"Thank you, Josephine, show them in please." Frederick Patrick remembered that he had once questioned how a parish, almost exclusively Polish, had been named St. Patrick's. The mystery vanished when he was told its pastor

17

founder had been a member of the FBI, the *foreign born Irish,* and not one to consult with parishioners on a matter of such importance.

The couple entered. The man made a head bow, the woman, an awkward curtsy. Frederick Patrick shook their hands. He was gracious as he seated them on the two leather padded client chairs that faced his desk. He walked back around the desk to his chair. "I remember your St. Patrick's parish well," he said, "from my visit there for confirmation."

"Yah, you confirmed our boy, Tommy, Bishop, two years ago," the woman answered timidly.

The bishop looked from wife to husband. "Tell me now, how can I be of service to you good people?" He tried to put them at ease, knowing that many people experienced their discomfort when ushered into a bishop's presence. "Mr. Krakowski?"

The man blushed, shifted a cud from one cheek to the other, looked at his wife, then lowered his head.

A moment of hesitation. "Father Crocker did things to our boy," the woman blurted, then stopped, her face reddening. The eyes of both parents teared.

"I'm not sure I understand you," the bishop said in a sympathetic voice. "Can you tell me what this is all about?"

Mrs. Krakowski continued, less timidly now. "Our boy, Tommy, has changed. He was always bright and laughing. He was always good in school. Two months ago he changed. He don't talk no more. He don't laugh. He don't do his homework. He gets into trouble. We were getting worried, you know. So one night we set him down and asked him what was going on. At first he says there was nothing the matter. But we pestered him. Finally, he comes out with it. He says the priest touched him. Our boy is out of grade school for three years now and goes to the public high school. But he still serves the Mass on weekends. After Mass one day, the priest tells our Tommy to come over to the rectory. We always taught our boy to obey the priest and sisters whenever they ask something. The priest asked Tommy to sit next to him on the sofa. He took the boy's pants down, and, and..." She stopped and looked down at her shaking hands.

"Bishop, we was always good Catholics," she continued after gathering herself. "And we work at the Church a lot. And we was always generous with money. But we never heard of no such thing as this. Did we do something wrong? What's to come of our boy? We don't know what to do," she pleaded.

"That's what the boy told you?" Bishop Sweeney asked quietly.

They nodded.

"Did you talk to Father Crocker?"

They shook their heads, no.

The bishop kept silent for a time. In the back of his mind he pictured Father John Crocker in this very office only a year earlier. Crocker had tearfully denied allegations of sexual abuse and claimed a smear campaign by disgruntled parishioners. The parishioners had offered only suspicions and allegations, not proof. It had come down to their word vs. Crocker's. The bishop had believed Crocker, but had reassigned him to Posnan "for a fresh start." He remembered the hard line he had taken against those parishioners. Now, he hesitated. "I'm sure you both understand that the Church must proceed cautiously with charges such as these. There are rules and guidelines we must follow here."

The bishop studied the couple. "I can see that you are very good people. And Father Crocker has always shown himself to be a good priest." Frederick Patrick's voice was slow and deliberate. "You must understand..."

The eyes of both parents were fixed on the bishop's face. Mrs. Krakowski's hands were clasped tightly in her lap. Mr. Krakowski's hands lay limp on his knees.

The bishop hesitated. "I want to commend you both. You did the right thing in coming to me. And I promise you that I will handle this and be fair to everyone. Your son has his rights, but so does Father Crocker. I will honor those rights on both sides." He paused, fingers massaging his pectoral cross. "And I wouldn't worry too much about your son," he said softly. "A young man like that has the ability to bounce back. I'm sure you good people have had difficult times in your life, and you bounced back."

Mrs. Krakowski shifted her weight toward her husband. Mr. Krakowski remained limp. Both stared at the bishop.

After letting the weight of silence burden the office for a time the bishop continued. "I know that you good people don't want harm to come to the Church. As a bishop I must do everything I can to prevent that. That's my job. You understand, I'm sure. We must not let this matter get out. So please understand that I must forbid you to talk about this to anyone but myself."

Frederick Patrick paused, studying the couple for their reaction. They gave none except for the tears that clung to their eyes. Mrs. Krakowski wiped her eyes with the back of her hand, and continued to stare at the bishop.

He began to gently paint a picture of all the forces of evil who were trying to destroy the Church: atheists, abortionists, euthanasia promoters, purveyors of sexual filth. These evildoers hovered around the Church like storm clouds, just waiting for someone to raise a target for their lightning. "I know that you good people don't want to do anything that would help such evil people. Certainly, you don't want to be the ones who give them ammunition to fire at God's Church. You will be doing God's will by leaving this matter in His hands. You can trust that He will take care of your son, never fear. Will you do that?"

They nodded haltingly, yes.

Frederick Patrick observed their reaction closely. "I want to thank you for coming to me with your story. You showed good sense in coming. And I want you to feel free to come back and talk to me again if you feel the need." He got up and, coming around his desk, shook their hands.

The audience was over. The couple stumbled out. The man stubbed the low sill at the office door and nearly fell. The woman held the man erect. Josephine McGregor watched them leave the chancery. They crossed the parking lot to their car and got in. The woman drove. They looked drained and were mute as they turned onto the road.

Frederick Patrick expelled a long breath. He sat for a moment and then reached for the phone. "Josephine, please call Father Crocker at St. Patrick's in Posnan and tell him I want to see him in my office at 10:00 tomorrow morning. And Josephine, do not mention the visit of Mr. and Mrs. Krakowski today."

CHAPTER FOUR

Eileen Fogarty waited in the hospital's lobby. She fidgeted with the arms of her wheelchair. Charles Mueller caught the brunt of her anxiety as he approached her.

"Father Mueller, get me out of here. Now!" she demanded.

Charles grinned. "Yes ma'am; right away, ma'am." He had received her call only ten minutes earlier. "What are you doing here anyway?"

"I had pains." Eileen clipped the words.

"Pains?"

"Chest pains."

"When?"

"Day before yesterday. Can we leave now?"

"How did you get here?"

"Josephine brought me."

He hadn't been called, but this wasn't the time to make it an issue. "Where's Josephine?"

"She's at work. Everybody's at work. That's why I called you."

Charles eyes danced. "You mean I don't work?" he teased.

"You know what I mean. Let's go! I can't stand this place."

Charles wheeled her out the main lobby door and down the ramp to his car, parked near the entrance in a spot reserved for clergy. He opened the car door, pushed the chair to it, and watched Eileen do a neat stand and entry. He folded the chair and put it in the trunk.

"Tell me about the chest pains," he prodded as they drove toward the college.

"I tell you, Father, I've been poked and pummeled, invaded and excavated from both ends, blood drawn, wired to crazy machines until I'm sick of it. I hate those doctors. What do they know?"

"The chest pains?" Charles persisted.

Eileen looked over at Charles. Sheepishly, "This goes under the seal of confession, now, Father?"

Charles nodded.

"It took them two days to discover it was only pressure from indigestion."

"You mean gas, don't you?"

Eileen started a raspy laugh.

Charles joined in. "Gas I can believe. I"ve certainly had my share from that Celtic mouth of yours."

Soon Eileen was laughing so hard she was catching her breath and swabbing tears at the same time.

Charles let her laughter funnel out the anxiety. He drove through the college campus and pulled into Eileen's driveway, directly across from the college's administration building. The transfer from the car, up the ramp and into the small white bungalow went smoothly. .

"Now, you just have a seat there, Father. I'll get us a little something to celebrate my good health." Eileen pointed at a chair.

"Tell me Eileen, did they check the octane in your gas? I'll bet that foine Irish whiskey of yours keeps it high enough to run a foine Cadillac."

"None of your smart lip now, Father. You know very well that I only keep the whiskey around to soothe the clergy, like yourself now." There was mischief in her eyes. "And you must work a little harder on that brogue. There's nothing harder on the ear than listening to Irish sung by a German choir." She wheeled over to Charles and handed him an album. "Here," she said, "I was looking at this when the pains came. You look at it while I tend to business." She turned her chair to a sideboard. Her hands quivered yet from the stroke that had disabled her, but she poured neat portions of the Jameson.

Charles did not offer to help. This was her ritual and she was its high priestess. He took his seat and opened the album. He looked at photos as his fingers gently massaged a leaf of the shamrock plant potted on a stand at his right hand. The room held the smell of earth and green plants.

Eileen, holding two cordial glasses on a tray in her lap, wheeled her chair to face him. She handed a glass to Charles. "Here's to the clergy, high and low." She held out her glass.

"Popes poop too, just so you know." Their glasses clicked and the ritual concluded.

"How old was Timothy in this picture?" Charles asked, holding up the album.

"He was forty-four. That was taken on our 24th anniversary, almost thirty years ago. My Timothy was dead the very next day." Eileen turned her face toward a corner curio cabinet. "You see all that Waterford and Belleek in the cabinet there? One piece from my Timothy on every anniversary. He never forgot."

"Eileen, if Ireland could be reduced to a single room, this would be it," Charles responded.

Eileen's crannied cheeks glowed with pleasure.

"Whose picture is this?" Charles held up the album toward Eileen and pointed.

"His name is Thomas Flanagan. Retired in Florida he is now. He was a close friend of Timothy, and was ever so helpful after Timothy died. As the months went on, though, he tried to get too friendly, if you get my drift. When I told him I was Mrs. Timothy Fogarty and I was going to die Mrs. Timothy Fogarty, he soon stopped coming around."

"I can just hear you saying it, Eileen," Charles said. "Not that you're overly independent, of course." She had told him once that a choice between purgatory and a retirement home was no choice at all. She pays a price for her independence, he thought. Not that she ever pays tribute to loneliness, or the wheelchair for that matter.

Eileen squinted at him. "You think you know me, do you young man? You look like you've been on Irish cooking. You tell Angela to put some weight on you. I'll not be painting your portrait with you looking like a scarecrow, or like some emaciated saint, which by the way, you're not."

"Instead of a portrait, how about painting the Holy Spirit for me?"

"Sure now, what does she look like, for me to be painting her?"

"How about a soaring eagle? I don't see the Holy Spirit as a dove. She's a love that soars above our understanding."

"I'm glad you reminded me. I've been meaning to talk to you about the preaching" She shook her finger at his face.

"Some folks aren't used to this loving, vanilla God you keep talking about. They're used to a persnickety God who keeps checking and knows if they've been naughty or nice. That way they come out of confession with that clean feeling, like they've had a gritty scrubbing down. Now, I know folks, young man, and I've been around a lot more years than you. So you just throw in some hell and damnation now and then. Let those people know that you're the boss and they'd better listen up."

Charles's right eyebrow lifted. "There's not many folks around who'll tell a priest that his sermons smell, Eileen. I appreciate your advice. And, if you get those pains again, and I don't hear about it right away, you're going to hear some of that hell and damnation you like so much."

"I knew you'd change the subject." The rasp in her voice carried into her laughter. "You're stubborn. You'd think you were Irish. But you just think about what I said, young man. You're heading for trouble if you don't get that preaching back in tune."

Twenty-five years in the chancery office as the former Bishop Zimmerman's secretary had left precious little Eileen didn't know about the clergy. Charles saw her concern. And it puzzled him. She had given him good advice during his stint in the chancery.

They took their whiskey slowly, the empty glasses a silent alert that the visit had come to an end. Before he left, Charles moved Eileen's easel to the window. Eileen had promised to finish Josephine's portrait for her birthday.

*　*　*　*

27

Josephine McGregor took dictation. She was good at it. It was the dictation that got her the job as Bishop Sweeney's secretary when Eileen Fogarty retired. Josephine typed the last envelope and attached it to the letter with a paper clip. She pulled a hand mirror from a desk drawer, checked her face and gave flouncing pats to the underside of her tightly coiffed *Desert Gold* bun. Then she picked up a finished pile of correspondence and walked into the bishop's office. "These are ready for your signature, Bishop. Shall I wait?"

"Yes, if you don't mind, Josephine. It shouldn't take long." Frederick Patrick Sweeney read and signed the letters. Josephine sat. Her fidgety hands first checked the brooch that sealed her blouse at the throat and then began a repetitive smoothing of her skirt over skinny thighs and knees. Peering through the thick lenses of round spectacles she studied the bishop's face. She liked the strong jaw, the red-tinged sandy hair, and the ruddy complexion. Such a handsome man, she thought, and then flushed at the audacious thought.

"Perfect as always, Josephine." He handed the stack back to her. "Thank you." A big smile compounded the reward.

"You're welcome, bishop," Josephine replied, relaxing now that he had dispensed her daily affirmation tablet. She took the mail and left the office. What a kind man, she thought, so appreciative.

Josephine knew from the first minute she laid eyes on you if you were worth the time or not. She called it her *gift.* "It's chemistry all right, but it's more than that," she had explained to friends. "It's like God is telling me." She never doubted her gift. If she got negative vibes on her first encounter with you, you couldn't market yourself to Josephine "nohow, nosireesir." If she did like you on that first meeting, you'd have to murder a nun to fall from grace with Josephine. She had been given the gift "to keep me safe." Josephine had been

28

safe for fifty-eight years. An only child who had been left financially comfortable, she was single and dedicated to the Church. She was sure that she saw people the way God did. She knew who was good and who was trash. Bishop Frederick Patrick Sweeney was wonderful.

Bishop Sweeney's eyes swept his office. He had made do with temporary quarters while the office had been completely renovated. It's my office now, he thought. The smell of fresh paint lingered. He stood and walked to the mahogany mantel above the red brick fireplace. There he studied the photos that crossed its surface. In one of them he stood, a newly consecrated bishop, between his parents. His mother, Hilda, canonized by the members of St. Bridget's parish for her giving nature, smiled out at him. That smile captured all that had made him a man at home with himself. His father, Patrick, smiled his Chicago ward politician smile.

"Are you busy, Bishop?" Monsignor Michael Higgins asked from the doorway.

"Not at the moment, Mike. Come in." The bishop walked over and met Higgins in the middle of the office. The six- foot two-inch bishop looked down at his red-faced chancellor.

Higgins had been a *wanna be* for the bishop's job when the former bishop, Bishop Zimmerman, died. He had campaigned effectively for popularity with the priests of the diocese. Unfortunately for Higgins, the Vatican had ignored the priests' recommendations. Frederick Patrick had obtained the position by gathering a higher constituency. The priests, instead of getting one of their own, got him, a Chicago export, but time and good marketing had healed their initial hostility. The bishop had replaced Mueller with Higgins, and Higgins had come around.

"We were going to discuss the Mueller incident, Bishop."

"I don't think that's necessary, Mike. I've thought about it. I'm meeting with Mueller on another matter today and don't intend to bring up the birth control thing. A lot of bishops aren't happy with the Vatican position. I just don't want *public* dissent. I think Father Mueller got the message. Don't you agree?"

"I hope he did," Higgins replied.

"I would be concerned, however, if Mueller taught his opinion openly at Padua College. Could you find a way to check on that?"

"Can do." Higgins left the office.

The bishop walked back to the fireplace and resumed his nostalgic journey. A second photo showed him at age 22, standing between his two grandfathers: Frederick Holzwerk, barrel-bellied, fun-loving master woodcarver, and Patrick Sweeney, leprechaun-faced, hard-drinking union attorney. He had been named for the two men, his first name coming from the German on a coin toss between them. It had been Hilda's law thereafter that he be called Frederick Patrick, never just Frederick, Fred or Freddie, and never just Patrick, Pat or Paddie. Hilda felt two names were prophecy of future dignity. The lawyer Sweeney regularly violated this law, but he was the only one outside of Frederick Patrick's own gang. The photo was taken on the day Frederick Patrick departed for Rome, where he had been assigned for theological studies. He had left with $500 in pin money from his woodcarver grandfather. From the other grandfather, the wealthier attorney, he had departed with advice. "Remember where you come from now, Paddie. Never fart higher than your own arse."

The bishop started to pace the office. The Krakowski visit had left him with mixed feelings. He felt himself at one with the Pope and his episcopal peers, and with their predecessors. Many had hurled threats of damnation to emperors and peasants alike. He had been more subtle. And he knew that it worked. Those simple people would never break silence. A nagging feeling of anxiety tried to displace the feeling of power. Lord, the boy is in Your hands, he prayed. Take care of him as You do Your Church.

Josephine entered the office and closed the door. "Father Mueller is here for his appointment, Bishop." She turned to go, hesitated, then turned back to the bishop. "You know how I hate to repeat gossip," she said in a hushed tone, "but I've heard that Father Mueller is quite a gambler. I thought you should know."

The bishop nodded.

CHAPTER FIVE

Bishop Sweeney looked around the table at the members of his presbyteral council. He felt an obligation to consult them, but none to follow their advice. They were, however, an invaluable link to financial support when it was needed.

"Reverend Fathers, I have one last item for our agenda today. It is the problem of the cathedral."

Michael Higgins jumped in. "No problem a good fire wouldn't cure, eh Bishop? Just kidding, of course."

"What is the problem with the cathedral, Bishop?" Monsignor August Oberkirche, 'Frosty' to his friends, joined in. "It is a beautiful church. It is old, ya, but the windows are so beautiful, blues like you see at Chartres. If repairs are needed, that is no problem."

The bishop made a sweeping gesture. "The entire structure is beyond repair in my judgment, Monsignor. It should be demolished and replaced."

Frosty saw the quickly stifled alarm in the eyes of several. "Ya, but have you had it inspected by the experts, Bishop, to see if it can be repaired?"

"It is the opinion of many, Monsignor, that repair is not the only consideration here. The cathedral is much too small for our times. Frankly, it is an embarrassment to me. Many of us feel that the Catholic presence in Mill Valley should be witnessed by a much more imposing structure."

A priest from a rural parish timidly braved: "Many of us were ordained in the present cathedral, Bishop. There is much sentiment for it..."

"I understand the sentiment, Father," the bishop cut in. "Unfortunately, there is a price to pay if we would give the Lord the building that He deserves."

"I'm with the bishop on this," boomed Nick Deutschman, pastor of the affluent St. Martha's parish in Mill Valley. "I say, go for it."

Nick was followed by an assortment of ayes, yeses, and nods around the table.

"Thank you, gentlemen," Frederick Patrick responded. "I certainly don't want to impose my will here. What we do is for the Lord's own greater honor and glory. Is there any further comment?" His eyes checked each council member. He knew the count without asking. Eight to four. The pastors of affluent city parishes were with him. The questioners were all rural. "If there is no further comment, and I do feel that I have the tenor of your feelings, I will let you know my decision within the next few weeks."

Higgins, the chancellor, cued in. "How will we pay for the new cathedral if you do decide to go ahead, Bishop?"

"I almost forgot that, Mike. Thanks for the reminder. The financing should also go before this council for your advice. It would be my recommendation that we impose a moderate tax on each parish of the diocese, based on their ability to pay, of course."

"How else would you get the money, Bishop? Print it?" Nick Deutschman's laugh drew support with nods and chuckles.

Frosty spoke out. "Should not the cathedral parishioners bear the greater burden, Bishop? After all, it is their parish. It is already difficult for many of our country parishes to support their priests and schools."

"Rest assured, Monsignor," the bishop conciliated, "we will take all of that into consideration. Are there any further questions or comments?"

On the way out of the chancery after the meeting, Monsignor John Haggett, rector of the cathedral, walked beside Frosty. "I hear you are planning to retire soon, Frosty. Will you stay in Burnside?"

"Where did you hear such a thing, John?" Frosty was startled. "I will retire when they carry me out in a pine box. I love my Burnside."

Haggett mumbled a reply and walked away.

Alone in his office, the bishop looked at the crucifix on the wall. Your Church in Mill Valley will have a new cathedral, Lord, he prayed. All for Your greater honor and glory.

* * * *

Charles Mueller entered the college administration building and turned toward the theology department. He had been alloted a small private office for counseling.

"Have you a minute, Father?"

Charles turned to face Sandra Wright, one of his students, from a wealthy Mill Valley family.

"Of course, Sandra, anytime. Come in." He led the way and closed the door behind them. Seated, he studied the troubled face of the pretty young woman. He waited.

Sandra was hesitant. "I don't know how to begin. I haven't talked about this to anyone, not even my mother, and...and certainly not my father." She looked into Charles's

35

eyes, then looked down. "I've told my parents all about your class, what you teach, and how much I like it."

Charles sensed the evasion. "Is my class what you want to talk about?"

"No. I want to talk about...I...I think I may be pregnant."

Charles waited.

"Jonathan...I mean...I...I don't feel attracted to boys. I...I thought maybe if I did it..."

Charles knew she was checking his eyes for any sign of alarm or judgment. "Is that it?" he asked kindly.

She nodded.

"You think you may be pregnant, and you're not sure about your sexuality?"

"Yes. I..." Sandra stopped.

"Did you get your answer about your sexuality?"

"No. I...I'm confused." Sandra's head was down, eyes on the floor.

"Look at me, Sandra."

She raised her head.

Charles waited for eye contact, then spoke slowly and tenderly. "Sandra, you are an attractive, talented, popular young woman. You are God's creature, loved by Him, valued by Him as much as anyone in this world. You are who you are. Do you understand what I'm saying?"

"I think so."

"I have a favorite saying, Sandra. It was found in 1692 in a Baltimore church. 'Be gentle with yourself. You are a child of the universe, no less than the trees and the stars; you have a right to be here'." Charles paused.

Sandra smiled. "Thank you, Father."

"Promise you'll remember that?" Charles smiled back.

"Yes, thank you."

"Now, Sandra, I think we should have you checked to see if you really are pregnant. And I think it would be good to have you checked on your *confusion.* Sound OK to you?"

Sandra nodded. "But, how?"

"I'm not the expert, Sandra, so I would refer you to a professional psychiatrist for that part. I have worked with her many times before. I don't know if a psychiatrist will handle the pregnancy part. If not, she'll make a confidential referral for that. And no one need know about this but you, me and the doctor." Charles paused and waited again for eye contact. "If it's OK with you, Sandra, I'll set it up for you right now."

"Yes, please."

Charles made the phone call. "Let me know how it turns out, Sandra. I'll say a prayer."

"I will, Father, thanks." Sandra seemed reluctant to leave.

Charles checked his daily list as he drove slowly down Main Street aimed for the Handy Man's Hardware Store. Angela had asked him to pick up some 100-watt light bulbs. He spotted George Schott as he came near the Hanging Tree Saloon. The saloon was named for an infamous oak, its woodpecker-drilled remnant still standing on the boulevard. Charles pulled to the curb and watched George sweep the sidewalk. He rolled down the window. "Hey George. You got a license for that kind of work?"

George looked up. Bloodshot eyes dominated his narrow face, as the face did his drooping frame. His smile was an effort. "Hi ya, Father Charlie." George came round to the passenger side, propped the broom on the fender, and climbed into the car. "You're just in time. I need a break." He wiped his brow.

Charles breathed in the stale whiskey smell. "I've seen you looking better, George. You been eating regular?"

George waffled. "Here and there, Father Charlie. You know. Say did you hear the one about the priest, the rabbi, and the leprechaun?" George started laughing hard and the joke never got started.

Charles had been through this before. He looked George over. "You have to eat, George. I've got a tab over at Harvey's Chicken Basket. I'm going to call Harvey as soon as I get back to the rectory and tell him you're covered."

George sounded a protest. "Naw, I..."

"Now, don't give me any guff, George. It's not a handout. You can work it off over at the rectory when you get the time. OK?"

George's eyes brightened. "Thanks, Father Charlie. It's been tight lately."

"Forget it. What are friends for, huh? Give me a call when you're free to do the rectory work. I'll ask Angela to make us a pan of lasagna. We can dig into that, have maybe a finger or two of Tony Corsalini's famous Dago Red, and shoot the bull a little. You can tell me that joke."

"Sounds good, Father Charlie. You get those jobs lined up for me. I'll be there. You can depend on it." George juggled the door handle and made his escape.

"See you, George." Charles knew it wouldn't happen. Even the lure of Tony Corsalini's homemade wine wouldn't bring George. But the invitation was out there. And he'd keep it alive, praying that it didn't outlive George.

Charles picked up the light bulbs and headed toward the rectory, radio tuned to country music. Eddy Arnold was singing *Make the World Go Away.* Charles reached into his jacket and popped a gumdrop into his mouth.

The black Buick parked in front of the rectory announced the presence of Monsignor August Peter Oberkirche. Charles parked in the garage and opened the kitchen door.

"Ya, Charles, how are you? It's good to see you." Frosty was sitting at the kitchen table with the housekeeper, Angela Pannetto. Open cookbooks covered the table, clear evidence that Frosty had been siphoning the talent of Mill Valley's best Italian cook.

"Monsignor, welcome! It's good to see you too." Charles held to the formal address, a measure of respect for the man who had been his pastor from birth until he had entered the seminary. "You're not trying to steal the best cook in the diocese, are you?"

The giant stood up, brown eyes merry. "No, Charles, you need Angela's cooking more than I do. His two bear paws grabbed Charles's outstretched hand and pumped. They stood, men of sharp contrasts, with open faced respect and affection. Frosty's bald head stood level with the top of Charles's shock. The hard barrel body of the older man equaled Charles's breadth at the shoulders. And Frosty's dark coloring contrasted the lightness of Charles's fair complexion.

"Business first, Monsignor. You will stay for supper and you will stay for the night. You get on that phone right now and call Gertrude. Tell her you will be home in time for Mass tomorrow. It's been too long since we talked. And Angela will make one of your favorites, won't you Angela?"

Angela stood and smoothed a rose-decorated apron over her tiny frame. Her waist seemed as small as a gallon jug and her warm brown eyes were in a constant melt. Flushed from the praise, she glowed. "You betcha, Father. Now you two get outta my kitchen quick or you ain't gonna get no supper." She shooed the two men out of the room and gathered the cookbooks from the table.

Charles sat while Frosty phoned the message to his housekeeper. He had been kidnapped but would be freed in the morning. Then Charles excused himself and made a quick call to Harvey's Chicken Basket.

"What are you doing now that the ice is out of the lake, Monsignor? It's too early for your garden. And what brings you to Mill Valley?" Charles knew Frosty's love for ice fishing and gardening. On most any winter day you could find him drilling holes in the ice with his big auger. He'd sit there by the hour, watching his tip-ups for the sign of a catch. No matter the temperature his ears stuck out from under an old biretta and his gloves stayed stuffed in his jacket pockets. Frosty had been tagged Frosty for his attachment to the sport. Late spring took him into the garden.

"It won't be long till I garden, Charles. Today, I had some business at the chancery. I will tell you about it later. But first, how long do I have to wait for a drink? Is this lent or something."

"What am I thinking?" Charles hurried into the kitchen and returned shortly with the makings. "Are you still drinking those brandy manhattans, Monsignor?" Knowing the answer, he was already adding bitters to ice.

40

"Ya, Charles, same as always." He watched as Charles, sleeves rolled up, blended ingredients. The long scar on Charles's forearm took him back to cinders and a boy with muscle in his soul. Charles had come riding his bike hell for leather to the church to serve Mass one day. When the boy wheeled onto the parish drive the tires slipped on freshly laid cinders. Frosty, who had been walking from the rectory to the church, watched as the boy's weight rode his forearm across the sharp jagged waste. Frosty slowed the bleeding by gripping the boy's upper arm while Gertrude drove them to the hospital. Then priest and housekeeper watched as the doctor plucked black broken spurs of cinder from the boy's shredded flesh and tied stitches to close the wounds. Charles's eyes wet involuntarily, but not a single whimper.

Drinks made, Charles took one first to Angela in the kitchen. Returning, he grabbed the remaining two, delivered one to Frosty and sat down. "Prosit!"

"Cheers, Charles!"

"How's Gertrude?"

"Good, Charles. She says you should come soon and she will make you that poppy seed cake you like."

"You be sure to give her my best." Each took a sip. "So, what's with this chancery visit today?" Charles asked.

"The bishop asked the priests' council in to tell us about his plans for the diocese. He is going to build a new cathedral, you know."

"No, I didn't know. Why?"

"We ask him that too, Charles. We tell him we like the old cathedral. It may need some repair, you know, but it is a wonderful old building. And the people love it. It has so much history. We tell him all that."

"What did he say to that?"

"He disagrees with us. He says the cathedral is a wreck and not worth repairing. He says he is embarrassed to celebrate Mass or conduct ordinations there."

"How is he going to pay for it?"

"He says he will make an assessment on every parish in the diocese."

"What are you going to do, Monsignor?"

"He is the boss, you know. I think it's dumb and I don't like it, but I do it." Frosty took another sip. "Charles, the bishop says something more what bothers me. Before our meeting started, he talked about those 'crazy disloyal theologians. They think they know more than the Church', he says." Frosty took a sizeable quaff, then another.

Charles matched Frosty sip for sip. Finally, "Did he mention any names?"

"He don't have to mention names. How many doctors of theology do we got in this diocese? We got only one," answering his own question, "and that's you." Frosty had felt proud that his boy had been selected to get a doctorate. "I'll bet you, Charles, you peeled a raw scab off his behind, didn't you. You been up to something to tick him off, Charles?"

"We had a little battle on the birth control issue. Nothing much. The whole thing will blow over in a few weeks. I was just saying what you would say."

Frosty beamed at the credit given, and capitulated. "Ya, Charles, you just be careful. This bishop is a tough bird." He used his index finger to swish the ice in his drink. "Now, tell me about Karl and Eva, how are they?"

"They're fine. My brother, Joe, runs the farm now. My folks built themselves a new home just down the road. Mom is ready to take it easy and go places. Dad wouldn't mind going places but is always too busy. When the weather's good, he's in the woods or on the bluffs hunting wild ginseng, or whatever."

"Ya, vell, you give them my regards when you see them."

The kitchen door opened. Angela announced supper. They would eat at the kitchen table, a practice favored by both priests. There would just be the three of them, Angela said. Father Bartell, the associate, had called. He was having supper with parishioners.

Frosty inhaled the kitchen smells. He was a man of smells: the heavy smoke smell from smoking fish or pheasants, the moist earth smell from the garden, the smell of incense on Sundays. People would choke, but Frosty regularly swung the billowing censer until the church fogged. Today he smelled like Irish Spring soap.

The men seated Angela and then took their chairs. Charles improvised grace. In front of them: a basket filled with slabs of homemade Italian bread, a large bowl of crisp garden salad, and, its odor steaming the room, a baking pan of red, bubbling lasagna.

Conversation was sparse, limited to acclaim absorbed by the contented Angela, and talk of ingredients. From time to time Angela would hop up to refill a dish. Frosty savored. Charles stopped eating when his stomach registered full.

The phone rang toward meal's end. Charles left the table to answer. When he returned he was putting on his collar. "That was Esther Worrell. An ambulance is taking John to the hospital. The paramedics say it looks like a heart attack." Charles grabbed a jacket from a hook by the door and put it on.

43

"Monsignor, if I don't see you tonight, I'll see you before you leave in the morning. You keep picking Angela's brain while I'm gone." He was out the door. It was nearly midnight when he returned to a quiet, darkened rectory. Frosty's caution about the bishop caught his mind as he prayed. He dismissed the thought.

CHAPTER SIX

"Howdy, ma'am. I'm here to check your gas meter."

"You're a rascal, that's what you are, Father Mueller. You just come in here and have some coffee." Eileen's raspy laugh followed her to the kitchen.

Charles followed and took the cup offered him. Back in the living room, he asked, "Any more chest pains?"

"Not a one, Father," Eileen answered. "But it's so nice of you to stop by. I don't think you've seen my painting of Maggie." She pointed to the wall.

Charles stood and walked closer to the painting. "No, I haven't." He examined the portrait. "You've caught her perfectly. But, now I'm confused. I thought you were painting Josephine."

"I am. I had this finished several months ago, and it's been hanging in my bedroom. Bill Smith moved it out here for me only yesterday. I decided I shouldn't hog it all to myself."

Charles felt riveted to the portrait. His eyes took in Maggie's every detail, the dark hair, aqua blue eyes, cream complexion. I'd swap a portrait of the Holy Spirit for this any day, he thought. "Eileen, you are first class. This is Maggie to a T."

Eileen beamed. She looked at the portrait. "My Maggie," she murmured. "Oh, did I ever tell you how I first met Maggie?"

Several times, Charles thought, but kept silent for the pleasure it gave Eileen.

"I was cleaning the house one late summer day and there was this knock on the door. I opened the door and there stood this beautiful girl, such a figure she had." Eileen paused at the visualization. "She looks at me and says: 'The dorms are full. Would you perhaps have a room for rent?' Well now, Father, I never had the notion to rent a room after Timothy died. But the look of that girl suddenly made it a good idea. Rent a room to her I did and she kept it those three years."

Charles watched the affection form on Eileen's face as she looked at the painting.

"If wishes could make it real, Father, she'd be my very own daughter. When she smiled she owned the world, she did, those flashing blue eyes and lovely white teeth. Boys coming round all the time and she sharing it all with me, such a good girl she was. I almost dropped my plate when she decided to become a nun. And she wasn't running from a broken romance. I'd have known that. What a mix I felt when she made her vows. And look now how far she's come! Doctor's degree, history professor, and now, president of the college."

Charles listened, eyes on the portrait.

"Still my Maggie she is, though," Eileen continued. "We do lunch every week. She'll be here tomorrow. I'll have chicken a la king on wheat toast."

* * * *

She profiled slim and feminine in sweater and jeans, and she held the reins softly. Her horse walked with an easy gait. They followed a trail that traced the curves of a stream roiling with the late spring melt. She wore a wide-brimmed hat that rode the flow of her dark hair. She absorbed the cool fresh air with deep breaths. Her eyes searched the early signs of spring. Willows were greening; pasqueflower, bloodroot and daffodils were breaking earth.

Sister Margaret McDonough, president of Padua College, · was taking a break from a school year nearly completed. I must come back in a few weeks, she thought. The spring scents are still missing. The blossoms will be full by then.

She crossed the boundary of the 5,500-acre Chippewa Wildlife Preserve. The trail led onto a path that wove between stream and pastures to a farmstead set on a rise. Maggie thought back to the Montana ranch of her youth, five brothers, no sisters, and all those tons of horses. She recalled the rivalry with her brothers. She had broken broncos, trained them and raced with her brothers. The body won't take that anymore, she thought.

She rode to the stable, dismounted, and looped the reins around the split log fence rail of the corral. She loosened the cinch, pulled off the saddle and hip carried it into the stable. Returning with a brush she began to curry the horse. "Thanks for the ride, Dusty," she said aloud. "What I mean is, thanks for the *easy* ride."

"So, just how sore is your butt, Maggie?" The voice came from behind her.

Maggie rubbed her behind. "I'll let you know tomorrow, Kate," she replied, laughing.

Kate carried Gretchen, her three-year-old, on her hip.

Maggie turned toward her perky blond friend and saw the child. "Hi, Gretchen. My goodness, did I tell you last night how big you're getting to be?"

Gretchen grinned. Her dark bangs and pigtails bobbed her pleasure.

Maggie led the groomed Dusty through the corral, removed the bridle and sent the horse into the pasture with a gentle whack on its hindquarter. She returned and reached out for the child.

"How about a horsey ride?"

"Yay, Yay," Gretchen shouted.

"Wheee!" Maggie swung the child onto her shoulders.

Gretchen screeched her glee. "Me ride Dusty!"

"No, you ride Maggie!" Maggie took off at a gallop around the corral.

Kate Brown and Maggie had been classmates. Together with Annie Johnson they had formed the terrible triad of the novitiate. They had survived on humor. On more than one occasion they had been severely reprimanded for "excessive hilarity" by the novice mistress, Sister Matilda Rufus, *Mother Attila* to the girls.

Once around the corral was all Maggie could handle. She wheezed up to Kate and swung Gretchen to the ground. "Horsey all tired out," she puffed to the little girl.

"More, more," Gretchen begged.

"That's enough for now, Gretchen," Kate said. "Maybe we'll ride Dusty this afternoon. Come on in, Maggie, the coffee pot's on. I saw you coming. Tony called. He said how good it was to have you with us last night."

"Tony's looking good."

"He's fine, works hard though, perhaps too hard." Kate Brown, a nun then, had fallen in love with Tony Corsalini, a priest, while on nursing duty at the hospital. Tony had been seriously hurt in a car accident. Maggie recalled Kate's early agonies over the relationship and her raw honesty. Maggie attended the wedding. Kate and Tony bought a 160-acre farm and converted it from dairy farm to livery stable. Now, Kate ran the stable during the week and did part-time nursing at the hospital on weekends.

The sweet smell of cinnamon filled the kitchen. Kate pointed to a cradle in one corner. "Karen is sleeping. I can't believe my good fortune with that one, not fussing all the time like you did, huh, Gretchen? Feed her and she's off to slumber town. A tornado wouldn't wake her."

Maggie went over and peeked at the sleeping baby. "Doesn't take after her mother, then, does she? Remember your night prowls around the convent? I can still see your face after Attila caught you red-handed at the fridge in the wee hours." Maggie's laugh was rich and musical. "Major infringement, that one! Seems to me you got duty on the laundry steam press for that one."

"Don't forget the times I saved your butt," Kate teased back. "How about the time I snuck up and locked Attila in the pantry so she wouldn't catch you in the kitchen. Sometimes I wonder how we ever made it through." Kate poured each a cup of coffee, and then brought a tray of freshly baked cinnamon rolls. They sat at the kitchen table. "We had so little time to talk last night, what with Gretchen demanding center stage. Bring me up to date. I read in the newspaper how you're getting famous giving talks all over the country, and from what I hear, you've turned Padua College into Camelot."

"It's exciting, Kate. I've been excited nonstop for the last four years. The talks I'm giving on the status of women are a sideline but good publicity for Padua. Padua is my main thing. The college is starting to get its own character. I feel lucky to be a part of it."

"Part, my foot. Charles told Tony that Padua is Maggieville. He said that the whole school has taken on your sparkle. How did he put it? He said, 'Maggie's blithe spirit has been institutionalized at Padua'."

Maggie tried to suppress a blush. "That was kind of Charles, but he was into hyperbole."

"Speaking of Charles," Kate said, getting serious, "if you don't mind my curiosity, is there anything going between you two? I've always thought you two are meant for each other."

"You mean romance, Kate?" Maggie also became serious.

"No. There's no romance between Charles and me. I still find all the romance I want in my Franciscan spirituality and in my job." She paused. "I admire Charles, though. No, that's an evasion. I like Charles, very much. There's no veneer on him. He's deep and direct, and able to throw light on the shadow side of things. And he's a terrific teacher. The students love him. The girls go gaga over him, and it totally escapes him."

"I think it's just that I'm so happy now," Kate said, "and I want the same for you."

"Not to worry, I'm happy." They talked for an hour before Maggie looked at her watch. "I've got to run, Kate. The rest of the day is full up. There's just time for me to shower and trade this horse odor for soap. Not everybody likes horse as much as I do." She stood and bent to Gretchen. "Maggie has to go home, Gretchen, do I get a hug?"

The child, sitting on the floor with her coloring books, jumped up and threw her arms around Maggie's neck.

"Oooh, such a strong hug; you are getting to be a big girl, Gretchen."

The child glowed. Kate stood and hugged Maggie. "Annie wrote," Kate said. "She'll be in town at the motherhouse for the summer. I'll want you both out here for at least a week if you can, so give me a call when you have your calendar."

"I'll do that, Kate. Bye Gretchen."

The two followed Maggie to her car. "Wave bye-bye, Gretchen," Kate prompted.

Back in the city, Maggie drove through the campus, enjoying the sight of it. Stone buildings dotted large spans of greening grass. Ancient oak and sugar maple broke the greenscape. It was true that the school had awakened from an academic snore. Maggie, however, was less generous to herself than were her friends in assessing the cause.

Maggie entered her office as the tower bell chimed one o'clock. Her appointment, Sister Felicity, was talking with Bill Smith, Maggie's secretary. "Hi, Bill. Anything that can't wait?"

"No earthquakes so far today, Sister," Bill replied.

"Good. Come on in, Sister Felicity." She led the way to a corner area of her office arranged for small group conversation. Four cushioned chairs surrounded a circular coffee table of rich cherry wood. A gentle fragrance from sliced applewood and flower petals rose from a small bowl in the table's center. Her desk occupied the opposite corner.

The lines of Sister Felicity's round face tumbled up to a fixed smile. She sat her considerable bulk into a chair. Maggie sat directly opposite.

"Something tells me you have good news, Felicity." Maggie felt profound respect for her director of student development.

Felicity's smile broadened. "The figures are just what you wanted, Sister Margaret."

"I can't wait to hear them."

"As of today we have accepted 1,624 students for the fall term. That's a nine percent increase over the current term's 1,490 students."

"That's wonderful." Maggie felt her excitement grow. "Why, that's over the target we set in our five-year plan. Great job, Felicity."

Felicity's smile widened. Praise from Sister Margaret was concentrated protein for her spirit. She would march through any hell but one for this boss. It had not always been so. "There's more," she said.

"More?"

"As of today, we have accepted 199 applications from foreign students." She consulted a note pad. "Forty-six from African countries, 59 from Central and South America, 12 from India, 36 from South East Asia, 27 from East Asia, and 19 from Australia. In all there are students from 22 countries."

Maggie did not try to hide her surprise. "Why, that is staggering; that's up more than triple. How did you ever do it?"

"Trade secret, Sister Margaret." The smile broadened to laughter.

Maggie's enthusiasm blossomed. "Felicity, those students will contribute more to the educational process than a dozen new professors. All those countries, all those cultures, open to all our students. That's so exciting! You did it, Felicity! You made it happen. Congratulations, and...and thank you." Maggie stopped and beamed her pleasure at Felicity.

Felicity used the chair arms to help boost her body to a standing position. Too much food for the soul was bringing discomfort. She faced Maggie. "I'm pleased that you're pleased, Sister Margaret. Now, I must get back to the office."

"Be sure to congratulate your staff for me, Felicity. And I'll come over later to do it personally." She walked Felicity to the door.

Maggie was standing at her desk when Bill announced the presence of Dr. Alexander. She motioned to send him in.

"Come in Dr. Alexander." Maggie walked to the seating area when Alexander appeared in the doorway. She motioned him over. "Please have a chair."

Dr. John Winslow Alexander walked over, a sway in his hips. A sea of gray-black hair waved and swirled on his head. Maggie stifled an expression of distaste as his cologne caught her nostrils.

"Thank you, Dr. McDonough." His voice was deep and nasal. He sat down, crossed his thin legs, and peered at Maggie out of a long thin face.

"The radio and newspaper publicity on your philanthropy to the Public Library was very complimentary, Dr. Alexander. Congratulations."

"Thank you, Dr. McDonough. It was nothing, really."

"I'd prefer that you call me, Sister, Dr. Alexander. You mentioned on the phone that you have been reflecting on ways that will enhance the prestige of Padua? I believe those were your words." Maggie gave him her full attention.

"Indeed yes, Dr. McDonough. Actually there is only one reflection that I am prepared to discuss today. More, perhaps, on another day. Moreover, we have already nuanced today's matter in previous discussions, but we've never hit it square on, so to speak."

"What matter is that, Dr. Alexander?"

"It's the matter of class load for senior tenured professors, Dr. McDonough. It seems to me that we are something less than civilized here at Padua in the classroom demands made on senior professors such as myself." Alexander sniffed.

"You mean the requirement that every professor, regardless of rank or tenure, must have a minimum of twelve hours per week classroom time?"

"Quite so. If Padua is ever to rise above the common it must as an institution begin to focus on the status of its professors in the academic community. It must promote research so that professors like myself can publish and become better known."

Maggie's eyes held on Alexander. Her arms rested easily on the arms of the chair. "We disagree, Dr. Alexander," she responded calmly. "Padua exists for students. All of our resources, including all professional energies will be focused in that direction, and that direction only. Padua is rising above the common, as you put it, but is doing so because it is a student-centered institution." She paused. "Now, certainly, some professors may feel more inclined toward a larger university where research can be accommodated. I assure you, we certainly want to foster legitimate aspirations of that kind."

Alexander reached down, picked up the bowl of potpourri, raised it to his nose with a flourish, and sniffed. "I must tell you, Dr. McDonough, that you are losing the confidence of your faculty with this relentless focus on students. Don't your faculty members deserve consideration as well?" He put the bowl back on the table.

"Indeed they do, Dr. Alexander. They deserve every help this institution can give them to make them better teachers. I don't see how taking them out of the classroom works to that end. Why do you want to get out of the classroom, Dr. Alexander?"

"Why, to do research and publish, of course."

Maggie leaned forward. "Isn't it a fact, Dr. Alexander, that you are practically out of the classroom now?"

"What do you mean by that?"

"I understand, Dr. Alexander, that your present student load is twenty-eight students, and those twenty-eight are spread over four courses. Why only twenty-eight students, Dr. Alexander?"

"I believe it is because I'm so demanding that I draw only the best students."

"Is it true, Dr. Alexander, that every one of your students received a grade of A last semester?"

"Yes it is." He hesitated. "But, as I've said, I only draw the best students."

"That's not the case, Dr. Alexander." Maggie walked to a file cabinet, opened it and pulled out a file. She sat down and opened the file. "Our student survey and follow-up investigations indicate you draw only students who need an A to raise their grade point average, those who desire to get that

A with little or no effort. The fact is, you have been buying students with grades, and even that bribe is beginning to fail. You may read the results, if you wish." She put the file on the table in front of Alexander and leaned back in her chair.

Alexander looked dazed. He did not reach for the file.

Maggie continued. "I regret the timing of this discussion, Dr. Alexander. I had intended to begin individual conferences with department chairpersons next week and to share the survey results. My comments to you as chairman of the English department would have been the same then."

She stopped and looked at the momentarily stricken Alexander. "You have so much talent, Dr. Alexander. You should bring it to your classroom. Throw away those yellow notes you've used in your classes for decades, and put some energy and creativity into your teaching." She made her appeal and then returned to the matter-of-fact. "You are free, of course, to take whatever course you choose, but my course is clear. There will be no retirement for professors while they still draw a Padua salary. And, any professor who continues to neglect his teaching the way you have done, won't last here, tenured or not. The choice is yours."

Alexander, recovered, glared at her. "You do me an injustice, Dr. McDonough. This is an unwarranted personal assault." He stood, turned to leave, hesitated and turned back to face her. "I...I am deeply offended." He turned again and strode out of the office.

Maggie remained seated for a time, feeling her rapid heartbeat. She knew the futility of any attempt at desk work and walked to the door. "I need some air, Bill."

The fresh air and the signs of spring were not enough to take her mind from John Winslow Alexander. She recalled trustees' comments that indicated Alexander was a frequent guest in their homes. He was reputedly wealthy from inheritance, and he had artfully spun a reputation with the trustees for wit and brilliance. She walked across the street, knocked on Eileen Fogarty's door and walked in.

"Saw you coming, Maggie. You've got that determined look about you. You're either a hornet about to sting or you've already done the deed and are looking back on it. You sit now, and we'll have a cup o' tea." Eileen wheeled to the kitchen.

Maggie could hear the tap turn on, the flow of water to the kettle, and the kettle meet the stove top. She knew better than to offer help. She kicked off her shoes and brought her feet to rest on a needlepoint shamrock that charmed the surface of a small footstool. She closed her eyes and breathed deeply repeatedly, and brought her heartbeat back to normal.

They sat for a half hour, the two of them. Eileen was the mare whose presence calmed the colt after a storm. She was Maggie's complete confidant. When the tale was told, Eileen cautioned. "It runs against my grain to speak uncharitably, Maggie, but I've known John Winslow Alexander for a lot of years. You be on your guard, now, with that man. Sure and there's more to him than meets the eye, and it's not the milk of human kindness I'm speaking about."

"I'll do that, Eileen. And don't you worry about me. I feel much better and I can handle it. Thanks for listening."

"Well, you know where the pleasure lies in that now. Don't forget lunch, tomorrow."

Resuming her walk, Maggie turned a corner onto the street that held the basement headquarters of Felicity's group. Her eyes caught the multicolored shirt of a student walking just ahead. *Gay Rights* was stenciled in pastels onto the shirt. "Good afternoon, Jonathan."

His shoulder length blond hair swirled as the young man stopped and turned. "Hi, Sister Margaret. How ya doin?" Jonathan's gentle voice contrasted with his strong masculine features.

"I'm doing fine, Jonathan. I hope you're keeping that four-point average going."

"I'm workin' at it, Sister. Physics is a bear, but other than that I think I'm on target. How 'bout if I repeat those straight A's, you'll let me publish my article in the *Paduan* next fall?"

Maggie's mind turned to the article and its promotion of gays' rights and lifestyle. "Dr. Alexander vetoed it, Jonathan, and I agreed with him. If you want it published in the *Paduan* you're in for a complete rewrite. You know my objections." She said it kindly. Maggie knew also that at least a few female students had surrendered to Jonathan's personal sexuality. She wondered what sympathetic nerve had made him so aggressive in the cause of gay and lesbian understanding. He needs a cause, she thought. But where will it lead him?

"I'll work on it, Sister. Good talkin to you. I've got to run." He took off running. Maggie turned to enter the Development Office.

CHAPTER SEVEN

Monsignor Nicholas Deutschman had finished the 7:00 a.m. Sunday Mass and stood in the sacristy removing the layers of vestments. First he pulled the flowing green chasuble over his head and tossed it onto the sacristy wardrobe. The stole and cincture cord followed with little effort. Next he grasped the long white alb at his beltline and used his fingers to inch the lower half up around his round belly. With effort he hoisted the alb over his black crew cut, the Deutschman logo. The alb also went onto the vestment heap. Volunteers would hang the vestments later.

The two altar boys returned from extinguishing the candles on the altar, one elbowing the other in the ribs. Stifling laughter, they removed their cassocks.

"So, what are you boys up to today?" Nick's voice rumbled around the marble walls.

The boys shrugged their shoulders.

"Cat got your tongue? Tell me, Johnny. You gonna be a priest?"

"Ah, no Monsignor. I'm gonna be a cop like my dad."

"Cop, eh? Well, I guess you better be a cop, Johnny. I don't think you're tough enough to be a priest." Nick turned to the other boy. "How about you, Matthew? You gonna be a priest?"

"I don't know, Monsignor, maybe." The boy tried to avoid the squelch delivered to his friend.

"Maybe? How old are you, Matthew?"

"Twelve, Monsignor."

"Twelve, eh? Well, that's old enough to start getting serious about what you're gonna be someday. You think about it, Matthew. You just might be tough enough to be a priest."

"OK, Monsignor." The two boys made a rapid escape.

Unvested, Nick knelt on the sacristy prie-dieu, took out a worn missal and read the same post-communion prayers he had recited since childhood. Then he stood, and, as a matter of habit before turning out the church lights, walked out into the sanctuary to check if the church was empty. Josephine McGregor was kneeling on the sanctuary steps. Nick retreated to the sacristy and left the lights on.

On leaving the church, Nick met his associate pastor, Father Art Huron, arriving for the 8:30 Mass. "Good morning, Father."

"Good morning, Monsignor." The perfunctory greetings summed up their relationship. Nick had held a tight leash on his two previous associates "to help them through the rough spots." One had told Nick to go straight to hell, and the other said he'd rather be in hell. Both demanded reassignment by the chancery. Bishop Sweeney had assigned the experienced Art Huron, ex-army drill sergeant, to the parish. On the excuse of his burdens at the Diocesan Education Office, Nick had delegated nearly all parish responsibilities to Art. Then he tried to fasten the leash on Art.

"Get the hell out of my way, Nick," Art had responded. "You gave me the job. I'll do it, but I'll do it my way." Nick had backed off, and dreamed of a small one-man country parish. If only those jobs had the same class. Still, Nick reflected, Art runs a good parish and I get the credit for it.

Helen Wieder, the housekeeper, was setting his breakfast table as he entered the dining room. His spirits lightened as he attacked the eggs and pork links. He recalled his doctor's advice to cut fat intake and lose weight. The good Lord will take me when He wants, fat or no fat. Nick gave the last two pork links to Rudi, his stocky German short hair, who was always at his master's feet at meal time. Rudi vented his appreciation by breaking wind. The odor plagued Nick's nostrils, but the sound did not make it past his bad right ear. He glared at his self-effacing housekeeper. Mrs. Wieder, smarting impassively under the same offensive smell, cleared the table, unaware of her assigned culpability.

God had proclaimed Sunday a day of rest and Nick gave grave observance to the Lord's will. It was his habit to give the remainder of the morning to a long nap, followed by a careful reading of the Sunday newspaper, followed by careful attention to spectator sports. The paper read, he turned on the TV. The final round of the Byron Nelson Golf Classic was on. From March through October Nick played at least three eighteens per week at the country club, but never on Sunday lest he give the wrong impression.

The helicopter-borne television camera traced the famous golf course hole by hole. The manicured fairways and greens and the scores of players already on the course took all of Nick's attention. Nick talked to the action. "You dropped your shoulder, Dummy! How could you miss a two-foot putt? You guys call yourselves pros? Damnation, won't somebody get that doorbell?"

Then he remembered. Art never stayed at the rectory on Sunday afternoon and he had personally ruled that Mrs. Wieder was never to answer the door when a priest was in the house. He wasn't about to have the impression get out that some woman ran the parish. Nick opened the door to the anxious face of a young woman. "Good afternoon, Magdalen," Nick said, not unkindly.

"I'm sorry to trouble you, Monsignor, but I need to go to confession."

"Come in, please, Magdalen." Nick led the woman into his office. He positioned two chairs so that Magdalen would talk at his good left ear. Both sat down.

"Bless me, Father, I have sinned," Magdalen whispered. "It's five days since my last confession. I may have sinned against the sixth commandment by impure thoughts."

Silence prompted Nick to ask. "Is there anything more?"

"No, Father." Magdalen sat, head down, hands clasped in her lap.

"Are you sure it was sinful?" Nick asked. He stared at the wall in front of his chair.

"No, Father. I saw this ad in a magazine. It was a man in bikini underwear. There was this bulge. I prayed *Jesus, Mary, Joseph,* over and over. But later I had all these doubts. Was it serious matter? Did I take pleasure in it? Did I give full consent? I couldn't make the questions go away. So I decided to come to confession just to be safe."

Nick also wanted to be safe. He could not bring himself to tell Magdalen she had not sinned, that she was suffering from scruples, that her problem was not sin, but emotional, and she would profit more from a psychotherapist than a priest. Absolution provided the more expedient solution for Nick as well as the woman. He dispensed a penance of one Our Father and raised his arm in absolution. He was back at the TV set in under three minutes. Sunday went down so easily.

CHAPTER EIGHT

Josephine had stayed in church after mass for her morning devotions. She knelt first on the sanctuary step and prayed before the tall, blood-stained, thorn-crowned corpus mounted on a cross. The crucifix rose above an old wedding cake altar and vied with it for prominence. Finished with the first step of her pilgrimage, Josephine moved to a side altar where a blued Madonna held a child in her arms while she squashed a wiggling snake under her heel. The Madonna's face was sweet and expectant, waiting for Josephine's memorare. Next, in a court protocol that must be observed, she crossed to the other side altar to pray to the lily-bearing carpenter, St. Joseph. I depend on you to keep my house in good repair and free from storm damage, she prayed.

In a side chapel, off the nave's middle, was the church's patroness, St. Martha. Martha reigned from a wall pedestal, balconied above a battalion of vigil lights. It was Martha's turn. Some years earlier, Josephine had blended a special recipe for devotion to Martha. She had first drawn the image of a floor-scrubbing Martha. Then she had inscribed the ejaculation, *St. Martha, Patroness of Small Work, Pray for Us,* around the drawing's perimeter. A printer reproduced the drawing on white cotton patches and Josephine had stitched each patch to a brown cloth backing and, adding a brown string, produced a neck scapular. She had then produced a larger version of the scapular as a kitchen wall hanging. To win the kitchen support of Martha (no burned potatoes, no fallen cakes, and positive success with corn relish) one need only repreat the ejaculation twelve times per request.

The devotion failed to develop widespread appeal. Josephine's best friend, Eileen Fogarty, spurned it, even though to Josephine's way of thinking, Eileen's cooking could use the help. Back then Josephine's imagination had floated gleefully on the thought of all the good she could do with the money certain to come from the sale of scapulars and wall hangings. Even when the devotion had come to a standstill, entangled by the intricacies of marketing, Josephine remained stoically confident. She continued filling boxes with scapulars and hangings. It's up to you now, St. Martha, she had scolded the saint. If you want this, do some miracle, or something. She just knew the saint was playing hard to get.

Josephine checked the time. Time for a short stop at the Infant of Prague. Eileen will be expecting me. Brunch! She quickly did the twelve ejaculations to St. Martha, counting all fingers plus two, for Eileen's Sunday brunch.

* * * *

Eileen Fogarty watched as Josephine McGregor and Bill Smith moved around her kitchen. They assembled food onto plates. She had done the cooking and was content now to let them serve it. Eileen had cheated on the baking. A nearby bakery had delivered a walnut patista the day before. Eileen had removed the bakery wrapping and today the patista, warm from the oven, rested on one of Eileen's own baking tins. She knew that Josephine would challenge its origin by asking for the recipe.

Seated at the table, Eileen led grace. "Thank you, Lord, for these good friends. Hold us in the palm of your hand. And, thanks dear God for the food."

"So, how's our good bishop these days, Josephine? He hasn't made the newsprint in over a week now." Bill's question was the kick-off to Josephine's favorite talk game.

Josephine had a mouthful of patista and a quizzical look. She chewed her way down to where she could talk. "That wonderful man! He's so busy!"

"I hear he throws a mean party," Bill returned.

"I'm sure they're very genteel, but I've never been to one," Josephine responded.

"He likes a party, he does," Eileen joined in. "And you've been to one, Josephine. Remember the party he threw for me? I hadn't been his secretary six months when I retired. Still, he threw me that party."

Josephine looked from the patista to Eileen, and back again. "Wasn't it wonderful, though?"

"Bishop Sweeney had everybody in stitches the whole time. When he wants to, he can be a real smoothie, that one can." Eileen laughed at the remembrance.

Josephine helped herself to a second slice of the patista. "I must have the recipe for this, Eileen. It's delicious. What do you call it?"

"Walnut patista. I'm so pleased you like it, Josephine. And, of course, you can have the recipe."

* * * *

At nine that evening, the ringing brought Eileen Fogarty to the nearest phone at hand, the one in her kitchen.

"Eileen, this is Josephine."

"How nice of you to call, Josephine. Are you sewing?"

"No, I've been baking. I simply had to call you." She sounded troubled.

"Whatever is the matter, Josephine?"

"I used your recipe and made that walnut patista. And it doesn't taste a bit like yours. In fact, it tastes awful."

"Well, now, did you pray to St. Martha?" Eileen asked.

"No, I forgot."

"The sparkle in Eileen's eyes belied the sympathy in her voice. "Well, sure, and that's it then!"

CHAPTER NINE

On Monday morning Nick Deutschman met the week with thoughts of Monday afternoon golf. As he walked to his garage he stopped to watch the construction workers on his unfinished parish center. The sight of Tony Corsalini talking to the job foreman brought a flush to his neck. He edged his short rotund body into the Lincoln and backed out of the driveway. Minutes later he pulled up at St. Stephen's rectory. Nick took on his enemies face to face, aware that his Roman collar immunized him from contradiction as effectively as Rudi's collar immunized the dog from fleas. Nick took frequent relish in poking his face into the face of others. "Good morning, Charles," he greeted curtly.

"Hi, Nick. Come in." Charles led the way into the office off the front entrance. They sat at a small conference table. "Can I have Angela make you some coffee, Nick?"

"No thanks. I have two things to talk about, Charles."

"Shoot"

"Number one, you've been stealing my parishioners. That's got to stop."

"Not true, Nick," Charles responded calmly. "The only recruiting I do is in my own territory. I have never recruited in your territory."

"Don't kid me, Charles." Nick pulled a sheet of paper from inside his suit coat and handed it to Charles. "That's a list of several dozen families who were members of my parish and are now in yours."

Charles took the list. "That may be, Nick, but I never recruited them."

"You should refuse them. You should send them packing right back where they belong." Nick's neck reddened.

"I don't agree, Nick. It's the people's choice. I don't recruit them, but I'll never refuse anyone. Are you recruiting in my territory?"

"I don't stoop that low."

"Well, perhaps you're not aware that some of my parishioners have left here and joined your parish. In fact, Nick, it seems they're some of the more rich and famous. Did you turn Stanley Wright down when he came to your door?" Charles handed the list back to Nick without looking at it.

"I don't want to stretch the importance of this matter, Charles. People like Stanley Wright deserve singular attention. He does so much for the Church. I was talking about run-of-the-mill people."

"If any of my people are more comfortable in your church, that's where they should be, Nick, rich or not. By the way, I drove by your place yesterday. I see that Stanley Wright got the job on your new parish center."

"I like to give work to parishioners when I can."

Charles let a moment of silence give conclusion to Nick's first concern. Elbows on the table, he picked at an eyebrow. "You said there was a second matter, Nick?"

"You certainly put the Church in a poor light, Charles."

"How so, Nick?"

"The way you presented the clergy at the Inter-Faith meeting last week. You make it sound like the Lord didn't put us in charge of the Church."

"He didn't put us in charge, Nick. He told his apostles they were to be servants. The pope and bishops get dibs on cleaning toilets. We're next in line."

"Bull roar," Nick erupted. "I grew up in a Church where the priest was the boss. That's the way it should be!" Nick pulled out a handkerchief, honked, wiped his nose, and continued. "The Lord started an institution to bring folks to heaven and I work for that institution. And, by damnation, I am a monsignor in that institution."

"You don't think the Lord put the institution ahead of the people, do you, Nick?" Charles asked.

"Yes I do, and yes He did. We do see things differently, Charles. And you know where I stand. If you don't change, well, you're in for a fight. I'm not going to stand by and watch you tear us all down. Now, I've got to get down to the education office." Nick rose abruptly.

"You're right, Nick. We do see things differently. Maybe we should learn to live with our differences." Charles saw Nick to the door. "Thanks for stopping, Nick. You're a busy man."

"Yes I am! A very busy man! I'll see you again, Charles."

The diocesan education office occupied space on the second floor of the chancery building. As he entered the building, Nick encountered the bishop entering at the same time. "Good morning, Bishop."

"Good morning, Nicholas," the bishop responded heartily.

"Would you have a few minutes for me sometime this morning, Bishop?"

"Of course. Give me ten minutes to get organized, Nicholas, and come down."

Nick spent the better part of an hour relating details of the Inter-Faith meeting and of his conversation with Charles that morning. "It all strikes me as disloyal and subversive, Bishop, and I thought you should know."

"Thank you for the briefing, Nicholas. It was appropriate that you share it with me. You are indeed a loyal son of the Church and I will ponder your message." Frederick Patrick gave the disappointed Nick no clue of his own thought or of any action he might take.

Nick returned to his upstairs office. The adjoining office was shared by Sisters Ann Dempsey, Salvatorian, and Martha Wagner, Franciscan. Sister Ann typed the last detail of their agenda for the week, walked the paper to the copy machine and made copies. "Here you go, Martha," she said, handing over a copy. "It's your turn this week."

"I'll buy donuts if you do it, Ann," Martha pleaded, jokingly.

"No deal. Besides you're better at making him feel like he does something."

"Better get it done, then." Martha took the paper and walked to the next door office. A plaque on the door read in three lines:

Monsignor Nicholas Deutschman
Doctor of Educational Administration
Director

Martha knocked on the door.

"Come in," Nick boomed.

Martha walked in. "Here's our schedule for the week, Monsignor. It's pretty much self-explanatory," she added, hoping to head off interrogaation.

The ploy worked. "Thank you, Sister. Is there any need for my participation?"

"I think we can handle everything, Monsignor. We'll holler if an emergency comes up. You're so busy."

"Very well, then." He waited until Martha had closed the door and then turned to the mail. No burdens there, junk mail to the last piece. Nick sailed them individually into the waste basket. Then he pulled a three-ring binder from a desk drawer and took up his Monday morning ritual. For an hour he consulted the Wall Street Journal and updated the values of his portfolio. He found it a satisfying task. That done, he took golf balls and putter from a closet and spent the remainder of the morning putting balls across the carpet.

CHAPTER TEN

A slight wind had carried the cool night air over the warmer river and bottomlands and produced a dense, heavy fog. The small car crept along the old river road. The car's occupants, an elderly couple, sat rigidly, eyes fixed to the road. The husband, both hands stiffly on the steering wheel, edged the car along the road's center and used the white line for direction.

A mile away and coming toward the couple, a rusty old Cadillac, radio blaring, moved at a healthy clip. The young man, alone in the car, rested unsteady hands on the steering wheel and straddled the road's center line.

They crashed at an S turn in the road. The heavy Cadillac crushed the smaller car, throwing the old man and his wife onto the roadside.

Tony Corsalini, on his way home from a late night on the job, pulled to the side of the road and stopped his pickup directly behind the flashing lights of the ambulance. Sheriff's deputies had positioned their squad cars, lights flashing, a hundred yards on either side of the accident. They were now placing flares along the road, alert to any traffic sound. Paramedics hovered over the two still figures on the roadside. A young man wandered about unsteadily. Tony recognized the newspaper reporter who was trying to engage the young man in conversation. A radio blared rock music from the Cadillac whose nose was buried in the swamp.

Tony walked quickly to the paramedics and the prone figures. He recognized the victims, even through the blood and disfigurement. They were friends, neighbors, and members of the same parish as he and Kate. He raised his hand in absolution, first to the woman, then to her husband.

"If you are capable, I forgive you your sins, in the name of the Father and of the Son and of the Holy Spirit." He traced the sign of the cross in the air. Lights flashing around him blended with those from the red and white ambulance and blue squad cars.

The next morning, pictures of the accident scene dominated the front page of the *Mill Valley Journal.* The two cars were pictured, the one crushed and formless, the other nose down in the swamp. The accident victims were hidden behind the active paramedics. Tony Corsalini, arm raised in blessing, was prominent and clearly pictured. The headline read: *Couple Meet Foggy Deaths. Former Priest Gives Absolution.*

* * * *

Bill Smith looked across the booth at Eileen Fogarty. He sniffed the hot coffee. "Lord, I wish I could make coffee as good as Harvey's," he said. He pushed the last bite of a cinnamon roll into his mouth. His back rested on the wall while his long legs stretched sidesaddle in the booth.

"You eat too fast, Bill. How many times have I told you, you're going to end up with indigestion, an ulcer, or worse." Eileen nibbled an egg sandwich.

The morning newspaper was spread out between them. Bill folded the comic section. "All I need to know I get from the funnies."

Eileen handed the front section of the paper to Bill. "Look at this, Bill. What a terrible accident. That Tony Corsalini is such a nice man."

74

Bill took the paper and started to read. "I didn't know ex-priests could give the sacraments."

"Why of course they can, Bill. That's because they're not ex-priests. They're ex-clerics."

"I don't get it." Bill fixed his gray eyes on Eileen.

"Bishop Zimmerman explained it to me. Tony Corsalini was laicized. That means he was a cleric, but now he is a lay person like you and me. He's still a priest."

"I still don't get it," Bill persisted.

"As I understand it," Eileen replied, "in the official Catholic way of looking at people, there are two kinds: clerics and lay people. Clerics make up the institution. Ordinarily a priest has to be a cleric to administer sacraments, but not if there's an emergency, like that car accident."

"So, if I get a heart attack, Tony Corsalini can hear my confession, and it takes?"

"That's right."

"Well, I think I like that idea. You never know. One of those ex's just might be Johnny on the spot at the right time." Bill walked his fingers through his salt and pepper hair, cut just long enough to part.

"You're a nut!" Eileen took the paper back from Bill and read in silence. "Oh, there's an editorial here about the bishop. The editor is giving him kudos for the design of the new cathedral." Eileen peered at Bill over the paper's top. "He also talks appreciatively about the boon to area employment from all the construction jobs that will be needed."

"I think the bishop has an *edifice* complex. I like the old cathedral just the way it is." Bill's face held its usual

75

unreadable expression. He shrugged his indifference. "Not that my opinion will change things."

Eileen raised her eyebrows. "I don't know if that's fair to the bishop, Bill. My gracious, look at the time." She wrapped the remainder of her breakfast back into its styrofoam container. "You had better get me home or you'll be late for work." She slid to the edge of the booth, hoisted herself with the support of the table and waited for Bill to swing the wheelchair into place.

*   *   *   *

Frederick Patrick Sweeney sat at his desk absorbed in the newspaper. The chancellor, Monsignor Higgins, had alerted the bishop to the news piece. Higgins waited, seated in front of the bishop's desk. The bishop finished reading, looked again at the photos, and frowned.

"Is this the Corsalini who was a priest of this diocese?" he asked.

"Yes he is, Bishop. He worked in the Marriage Tribunal here at the chancery at the time he left. Charles Mueller was chancellor then and handled the paper work for Corsalini's dispensation."

"I've seen this Corsalini's name often in the newspaper for one civic project or another. I guess I never put two and two together. Why is he still living in this area?" The bishop stared at Higgins.

"I can't answer that, Bishop. I wasn't here at the time."

"Do we know what he does for a living?"

"I'm told he works for Wright Construction."

The bishop sat back in his chair. "Bring me his file, would you please, Mike?"

"Right away, Bishop." Higgins moved to the door.

CHAPTER ELEVEN

After seeing Eileen home and into her living room, Bill drove across the street and parked in the lot next to the Padua College administration center. What unexpected paths our lives take, he thought, staring up at the building. He mentally tracked his own recent history, his retirement as bank president when his wife died three years earlier, how he slowly exhausted the terrible grief by charting a new course, first by auditing several courses here at the college. Then, to fill time, he had volunteered as a fund raiser for the Padua Development Fund.

He recalled the infectious optimism that went with Sister Margaret's vision for the college. You hooked me, Sister Margaret, he thought. Before he knew it, his excitement matched hers. When her secretary resigned to get married, Bill, to his own surprise, had volunteered as a temporary support. He never left. Much more than a secretary, he was guard at her door, financial counselor and business confidant. He reinvested his salary in the college library. Bill smacked a fist into the open palm of his other hand. I'll be sixty-eight next month, and I never thought I'd ever have this much fun again. He whistled his way to his office.

Charles Mueller walked into Bill's office. "Hi, Bill. You've got me in the appointment book."

"She's in, Father, and she's not on the phone. Go right in."

Maggie, reading glasses on, was at her desk behind a pile of correspondence. She looked up at the movement at her door. "Hi, Sister. Is it convenient?" Charles asked.

"Come in, Charles. It's convenient." She rose from her desk and moved to the conversation area with him. She picked up a bowl from the table. "I know you have a weakness for gumdrops and gingersnaps. Will a lemon drop do?"

Charles took one. "That'll do just fine. Thanks. My only problem with lemon drops is I crunch them before they're all sucked down." He stuck the lemon drop in his teeth.

"I received a call yesterday from Bishop Sweeney," Maggie began. "He asked me to see him on what he called 'a matter of concern'. I asked him the point of his concern and he muttered vaguely that it had to do with the theology department here at Padua. He either could not or would not be more specific. We're scheduled for tomorrow morning. I'm in the dark here. Do you know what his problem is?"

Charles thought for a moment. "I'll bet his problem with the theology department is just a ruse. He wants to meet you personally and test your metal. The real problem will turn out to be your growing reputation as a radical feminist." His eyes shimmered.

"I am not a radical feminist, Charles Mueller," Maggie shot back, then paused. "But, I'll do until one comes along."

"I suppose you even think women have souls?"

Maggie laughed. "Women have common sense, Charles, so they must have souls. I wonder sometimes about men."

"Good. I can see that your metal won't melt under episcopal fire. Now where were we?"

"We" she stressed the word, "were talking about a problem, Charles, accent on problem, a problem the bishop sees in your department, Charles."

"Oh, that problem."

"If we can get down to serious without damage to your mental stability," she teased.

"OK, serious it is." He crunched the lemon drop and reached for a second. "The short answer is that I'm not aware of any problem. The long answer is long. If Frederick Patrick Sweeney sees a problem, it can only be one of two things. It's either a problem with the people in the department or a problem with what we teach. I doubt if it's a problem with the people. There's only three of us. Sister Rosalie has a solid theology background obtained from the Jesuits. And contrary to what you may think, that exotic dancer on the Peek A Boo Club's billboard is not Rosie."

"You were going to be serious?"

"Ah yes, serious. Jake Costello has a solid academic background from Notre Dame. Married, four kids, two of them at Padua. He's good, remember? Imagine a theology prof getting voted Teacher of the Year by the students as Jake did last year. He could get a job anywhere. But he stays at Padua because Padua gives free tuition to faculty kids." Charles paused.

"And there's you," Maggie interposed.

"Then there's me." Charles turned slightly to stretch his long legs along the edge of the coffee table. "You will never believe this, and I trust your discretion, but I have been a problem for Bishop Sweeney at times."

"Not only would I believe that, Charles, I would expect it. Are you the concern Bishop Sweeney is talking about?"

"We have differed at times on matters like birth control," Charles answered. "But, if I were the department's problem, he wouldn't consult you. He would fire me with a simple transfer, that age-old bureaucratic method of problem solving, anytime he wanted to."

"I see." Maggie's blue eyes studied Charles. The repartee camouflaged the fondness she felt. "So?"

"So, what?" Charles asked innocently.

"I presume this path of what the problem is not will take us to what the problem is, or at least might be?"

"There's no need for sarcasm now," Charles cautioned. "We are done with simple logic and now enter the world of speculation and fancy. Are you ready?"

"Ready. Be as fancy as you want."

At that moment, Bill entered the office. "I'm sorry to break in on you. Campus Security just called. There's a disturbance on campus and it's headed our way." The windows next to the conversation area permitted full view of a green space that extended nearly a block. At the far end of the green, two bodies ran toward them. They were naked bodies, one male, one female, both with mask and wig. Body parts bobbed and wiggled as they streaked by and disappeared.

Charles looked at Maggie. "Is that part of the new Physical Ed. curriculum, Sister Margaret?"

Maggie looked down and shook her head. "No, it's a disease that affects students at exam time." They laughed and Bill excused himself. Maggie and Charles sat down again and took a few moments to regain their composure. "Now, where were we?" Maggie asked. "Oh, yes, you were going to be fanciful."

"Right." Charles crunched the lemon drop. "My fancy pictures a bishop concerned with authority. It goes like this.

82

Who knows how, but some member of Bishop Sweeney's cocktail set tunes in to what we teach at Padua, and hears a distant sound from the catechism lessons of their grade and high school days. That person mentions this to the bishop. Frederick Patrick has a vision of rebellion growing in his own backyard. His Church might be threatened, that is, his authority might be endangered. I'm not trying to demean the bishop, Maggie. I respect him. He is a man true to his own vision of his Church, and of his role in that Church. I don't have that same vision. He will want reassurance from you that all's well, either at your meeting tomorrow or after you have conducted a holy inquisition."

"You don't mean that?"

"Maggie, I'll bet you the next cup of coffee that he is concerned that a virulent species of modern heresy is threatening his Church."

"What kind of heresy?"

"The one that pits the Church as an institution against the Church as a people of God and would make the former servant of the latter. And, Sister President, we are guilty."

"I don't see..." Maggie was interrupted by Bill.

Bill spoke from the doorway. "I'm sorry to interrupt again, Sister Margaret, but the people in Seattle need to talk to you about your speech. They say it's an emergency having to do with publicity."

"I'd better take that call. Do you mind, Charles?"

Charles stood and went to the window, but his vision was inward. The scene was Washington, D.C. He was a student. Bishop Zimmerman regaled him on church matters while inhaling a dozen oysters on the half shell. Zimmerman was attending a meeting of the Catholic bishops; the bishop referred to it as "the most exclusive club in the world". The description had stuck in Charles's mind. He wondered why it had taken so much time for him to comprehend the practical antipathy of those words to the gospel.

Another scene came to his mind. He was in the bishop's office here in Mill Valley. Charles was now chancellor to Bishop Zimmerman. "I understand the difference between us, Charles," the bishop was saying. "You think it is more important to be a Christian, and I think it is more important to be a Catholic." There could not be, Charles thought now, a more succinct or clear statement of their differences, the people of God vs. the clerical institution.

"Sorry about that, Charles." Maggie interrupted his thought.

"No problem." Charles sat down again, picked at an eyebrow and looked at Maggie.

"I understand what you said about the two churches, Charles, but aren't you making a bit much of this. After all, many church documents speak of the Church as the people of God."

"Yes, they do. I'm talking about how the institutional Church operates in the real world. And, Bishop Sweeney *is* the institution in our neck of the world. He is mentally and emotionally hooked to that Church. And he is smart. He is streetwise in his clerical world and can walk the alleys of that world with assurance. He can smell the scent of threat to the power of his Church. If I'm right about what he has in mind for your meeting, don't look for a debate on the issues. He will approach the matter with power and righteousness."

Both were silent for a time. Maggie spoke first. "Thank you, Charles. I am not a theologian, but I am not a stranger to power plays. Academic freedom seems constantly under seige. Someone is often at my door demanding that a particular book be burned, or a faculty member dismissed for possessing a contrary point of view. And you can bet that the streaking incident we've just witnessed will come back to haunt the college. Your perspective helps. Oh, I'll take you up on the coffee bet."

"Good. How's your lemon drop supply?" He reached for another.

"Why did you leave the chancery, Charles? Wouldn't that have been a better forum for you to make your points?" Maggie's periodic efforts to plumb the man increased her interest as more bits of his personal scrapbook fell into view.

"I was thinking about giving up the chancellor job when Bishop Zimmerman died so unexpectedly. Then I got fired by his successor, Bishop Sweeney."

"You did not get fired, Charles Mueller."

"Yes I did, Margaret McDonough. Frederick Patrick Sweeney advised me within days of his coming that he planned to make a clean sweep of chancery officials. 'I want to forge my own team,' he said. I had no problem with that. He was decent and straightforward about it. So he gave me St. Stephen's parish, the only plum in his bag at the time, as he put it."

"Is St. Stephen's a plum?"

"It's people, like any other parish, the full spectrum of human needs and troubles." Charles pushed himself up from his seat. "I'll be curious to find out how your meeting goes."

"I'll let you know." Maggie stood and walked Charles to the door. "And thanks."

CHAPTER TWELVE

Bill Smith worked the crossword puzzle. "What's a four-letter word for 'feed the pigs'? he asked. He took a sip of Harvey's famous coffee.

Eileen Fogarty's face appeared over the newspaper. "Slop? Maybe?"

"Slop? Good, slop works."

"My goodness, Bill, here's an article on Maggie. The headline reads *Padua President Promotes Poop (Power).*

"Oh, oh! Sounds like Sister Margaret. What's the article say?"

"Listen to this."

> *The president of Padua College, Sister Margaret McDonough, fielded a challenge from Monsignor Joseph Post of the National Conference of Catholic Bishops at the Religious Organization of Women's (ROW) Conference in Denver, Saturday.*

> *The exchange took place minutes after McDonough delivered the keynote address in which she stated that women are excluded from positions of power in the Catholic Church.*

> *Post questioned why women would even want to be priests. McDonough responded that you must be a priest to be a cleric, and that only clerics hold power in the Church.*

> *Post then questioned whether women need power in an organization founded on love.*

McDonough responded that love is not subservience and that power speaks equality.

In Post's final challenge he stated: "The exclusion of women from clerical status does not eliminate a woman's power or her equality. Power and equality are found in holiness, which is open to all."

"That's smoke, Monsignor. Exclusion means exclusion," McDonough replied. "Once you've been disemboweled, it is forever impossible to poop."

The conference, attended by over 700 religious women from the west and midwest, discussed the role of women religious in the Church. Post was an invited observer.

Prior to the above exchange, the Padua president had traced the role of women in the Church down the centuries. She gave numerous examples of the explicit misogyny of famous churchmen. She said, for example, that Albert the Great had claimed that women are misbegotten men with defective natures, and that Albert's student was Thomas Acquinas, whose influence greatly impacts the Church even today.

McDonough accused the American hierarchy of a conspiracy of silence with the Vatican. She charged that the bishops and the Vatican Curia have been unwilling to debate the empowerment of women in the Church with openness, or with women participants in the debate.

Eileen put the paper down. "Now, Bill, doesn't that sound just like Maggie?"

"It does, but I wish she hadn't said some of those things, or better, that the newspaper had left them out. Her comments were meant for an audience of women religious at a conference in Denver. I'm not sure how they will play with Bishop Sweeney. Sister Margaret has a meeting with him today."

CHAPTER THIRTEEN

Josephine ushered Sister Margaret McDonough into the bishop's office. Frederick Patrick Sweeney rose from his chair, came from behind his desk and welcomed her with a display of gracious attention. "Thank you for coming, Sister Margaret. I know you must have a busy schedule and I appreciate you giving me your time." He reached out to shake her hand. "Please have a seat." His arm swept toward one of the client chairs that faced his desk.

"What an unusual desk you have, Bishop." Maggie looked around the elegantly appointed office. "All of your furnishings are so unique." She caught the sweet scent of the cut flower arrangement on his desk corner.

"Why, thank you, Sister Margaret," the bishop responded. He smiled his appreciation. "The desk was used by my grandfather in his law firm and I have inherited it. It's solid walnut. The drawers actually reach through the desk and can be opened from either front or back. My other grandfather, on my mother's side, made the desk and hand carved the detail." He caressed the polished wood, then pointed to the intricate figurines carved into the legs of the desk.

The niceties were soon finished and Frederick Patrick leaned back in his chair, fingers of both hands fondling his pectoral cross. "Sister Margaret, what is your opinion of the competency of your theology faculty?" The words came out elongated and formal.

"I have a very high opinion of their competency as teachers, Bishop. I am not a theologian and so must rely on their academic backgrounds as a guide to their theological competency. I believe you have met each of them on one occasion or another. Certainly you know Father Charles

Mueller, the department chairperson. He is one of your priests." Maggie went on to review the academic backgrounds of Sister Rosalie, Jake Costello and Father Mueller. She also commented on the student assessment given each and on her personal character assessment of each, both highly complimentary. "Do you have a concern about any particular faculty person, Bishop?"

"For now, let's just say that I've been hearing things that disturb me, Sister Margaret, things that smack of disloyalty to the Church, things that," he leaned his head farther back, "in my judgment, militate against a solid Catholic identity for Padua College. I'm sure neither of us wants the college to have that identity compromised?" The words rolled out with precise diction.

Maggie picked up on his lead. "I would hope not, Bishop. I believe that the college's identity as Catholic and Christian emerges most clearly in the quality of education we give our students. In what way do you see our Catholic identity threatened?" Maggie's voice remained calm despite the alarm bells sounding in her head.

The bishop paused and maintained eye contact. "Let me reply by way of example. Last week I attended a meeting of the more important businessmen of the area. One of them told me that his daughter had come home quite enthused from her religion classes at Padua. Upon inquiry it turned out that the parents were less enthused. Whatever was discussed there did not seem Catholic to them. Now, Sister Margaret, these are important people in our community. According to the parent, there seemed to be some questioning of papal teaching by the professor. I did not go into greater detail with him, but I did assure him that I would look into the matter. Sister, this is not the first time I have received such comments." The bishop resumed stroking his pectoral cross, eyes on Maggie, expectant.

Maggie lent dignity to the bishop's words by her hesitation. Her facial muscles squiggled in concentration. "I'm wondering, bishop, if the student might lack the maturity or breadth to..."

The bishop interrupted. "I assure you, Sister, this student is very bright."

"But, for the parent to impugn either the competency or the integrity of a member of our faculty?"

"Sister Margaret," the bishop lowered his voice and spoke softly, "perhaps I have more experience in this area than you do. Vigilance in the matter of doctrinal integrity is my job. And where I have found smoke in the past, I have usually found fire."

Maggie leaned forward, cradled her knee with both hands, and asked. "But, do we know if this was a doctrinal matter, Bishop? You would agree, I believe, that there can be legitimate differences with the Vatican on non-doctrinal matters. Perhaps this was such a matter?"

"I do not agree, Sister. In my book doctrinal or not makes no difference. Catholics are obliged to give respectful hearing and obedience to all church teaching, even if it is not a defined dogma. One must have respect for the Vicar of Christ."

Maggie felt her heartbeat escalate. Was she about to walk off a cliff? She had fought this battle with herself before. The Maggie of her youth wanted to blurt it out straight and unvarnished. The professional Maggie, the diplomat Maggie, saw the need for varnish. "Bishop, I would respectfully make a distinction between deference and respect. I most certainly give deference to the Pope and the Vatican Curia. But I agree with Cardinal Newman that the first Vicar of Christ is my conscience..."

"Sister Margaret," the bishop interrupted, "I learned from this morning's newspaper of your disaffection with the hierarchy."

"I'm sorry, Bishop?" Maggie's face showed her puzzlement. "I did not see the morning paper."

"Apparently, Sister, you are gaining a reputation as a national spokeswoman for women's rights. There is a news article about your recent Denver speech. According to the article, you accuse the American hierarchy of a conspiracy, a *cowardly* conspiracy from the tone of the article, with the Vatican on the matter of women's rights in the Church. I must tell you, Sister, that I see the Church as a champion of those rights. Women have always been honored, put on a pedestal if you will, by the Church. Why, the Blessed Mother holds a place of honor next to Christ himself. You yourself hold an honored position. I see no conspiracy, and certainly not your implied cowardice, in the Church's hierarchy." His words were deliberate. "I assure you, Sister, I am no coward."

The age-old dilemma, Maggie thought. How do you say there are good people doing bad things, and the bishops may be some of those good people, without it getting personal? "I trust this is a discussion, Bishop. It is not something personal between us. Father Mueller has described you commendably as open and straightforward. I hope you will take my response as a sign of trust in that description of you. Now, as to the points you made: First of all, I have never used the word 'cowardice' in regard to the hierarchy. If the newspaper article implies that I did use that word, it is in error. Next, there simply is no history of appointments of women made by the hierarchy to positions such as I hold. I hold my position because the college is owned by the Sisters and appointments are made to the position by a board independent of the clergy.

"But, I'm not the issue here nor is the Blessed Mother. Though, I think sometimes the hierarchy gives such honor to Mary because, number one, she's dead, and number two, they have so emptied her of her humanity that she has become a disembodied abstraction. She is no longer seen as the real human mother of the real human Jesus, the nurturer and enabler of a man with sufficient character and courage to face the cross. Their focus is on her virginity as if that somehow validates honoring her."

Maggie stopped, and watched as the bishop inspected his fingernails. She tasted the humiliation of dismissal. She tried to open a conciliatory exchange. "Women need help today, Bishop, to overcome the barriers that this mentality has created in the Church. My goodness, we have no voice whatsoever in the Church. We need strong supporters in the hierarchy to help change this. I hope you will be one of them."

"Well, Sister Margaret, I appreciate your giving me your views. We must continue this interesting discussion another time. I've advised you of my concern regarding the orthodoxy of your theology staff. Please give the matter some observation. Certainly, academic freedom is a bit nebulous in this context."

"It's not nebulous at all, Bishop. One either has academic freedom or one does not..." Maggie saw the fingernail inspection resume. "However, I certainly don't mean to dismiss your concerns. I would be happy to have you express those concerns directly to the theology department staff. That way you would receive their response directly, rather than through my unqualified perspective or that of some student. You know also that Father Mueller is the chairperson of the department. He is under your immediate authority. May I set up a meeting for you with Father Mueller and the other members of his department?"

"I will consider that option, Sister Margaret. In the meantime I will anticipate your resolution of the problem. I'm very glad we had this opportunity to get to know each other better, Sister. Thank you for sharing your valuable time."

Maggie smiled her open and trusting smile.

Frederick Patrick rose to see her out. At the door, after Maggie's departure, he asked Josephine to have Father Curley see him. He was barely back in his chair when the young priest, red hair glinting under the bright light, entered.

"You wanted to see me, Bishop?" Curley was Frederick Patrick's handyman, chauffeur and master of ceremonies. When not so engaged, he worked in the Marriage Tribunal in the chancery.

"Yes, Father. It won't take a moment," a sign to the young man to remain standing. "I want you to get me the names and some biographical data on all members of the Board of Trustees at Padua College. Be discrete. I won't want my interest signaled in any way."

"Yes, Bishop, I understand." He left.

Independent of the clergy, are they? the bishop thought. We'll see. He called for Josephine, who was there in an instant, and began a letter to Cardinal della Tevere in the Roman Curia. The letter advised the cardinal of Frederick Patrick's coming visit to Rome and sought an appointment. As a kind of filler he included a note on Sister Margaret McDonough's feminism and on his suspicions regarding the theology taught at Padua College.

"I am confident of my ability to quell the disloyalty without undue publicity," he wrote. "I would, however, be most grateful if Your Eminence, with your years of experience and wisdom, has any advice for me in this matter. I am sometimes astonished that people, so otherwise talented, fail to see the depth of their disloyalty in espousals of this kind." He concluded, "I look forward with great delight to once again being renewed and strengthened in your presence."

Father John Curley was back in the bishop's office when the chancery opened the next morning. "The Board of Trustees at Padua College is composed of eighteen members, Bishop. Currently, there are three non-Catholics, eight Catholic laymen and women, five members of the Franciscan Order, and two vacancies. Curley listed them by name. Frederick Patrick knew most of them on a first-name basis. He knew Sister Mary Jane Plummer, President of the religious order, but not the other four nuns, and he had not yet met one of the non-Catholics. With the same exceptions, each represented a power in the community. "I have it in confidence, Bishop," Curley continued, "that Stanley Wright of the Wright Construction Company, a Catholic as you know, and Samuel Stein of Stein Manufacturing, a Jew, are the Boards' nominees to fill the two vacancies. The nominations went to the two individuals only yesterday so they have not yet confirmed their acceptance. Is there anything else you need, Bishop?"

"No, thank you, Father. Good work. That will be all for now."

Frederick Patrick was reaching for the phone as the door closed. "Please get me Stanley Wright on the phone, Josephine." A minute later Wright was on the phone. "Good morning, Stanley, I hope the construction business is booming," Frederick Patrick enthused.

"We're doing well, Your Excellency, thank you. I hope the Church is doing even better. How can I be of help this morning?"

"Stanley, I have it in confidence that you've been nominated to the Board of Trustees at Padua College. May I ask if you have given your decision yet?"

"Actually, Your Excellency, I have not yet received a formal notice of nomination. I have indicated a willingness to serve."

"Then the notice will probably arrive in today's mail," the bishop replied. "I'm very pleased, personally, that you are willing to add this burden to your heavy workload. Could we have lunch this week? There are several matters I would like to discuss with you in private."

CHAPTER FOURTEEN

The noon bell sounded the end of class. Charles Mueller collected the exams, joked with the late finishers, and walked out of the classroom. Headed toward the college cafeteria he retrieved a gingersnap from his briefcase and plunged it whole into his mouth. June had begun clear and warm. The scents of lilac and flowering crab mingled as they drifted through the open hall windows. Today was the last day of the spring semester. Charles had taught two classes of theology on both Tuesday and Thursday mornings. The classes added a burden to his workload at the parish, and preparation for them often meant late hours and loss of sleep. Still, he couldn't let them go. They kept his mind alive.

"Good morning, Father Mueller." Charles turned. It was Maggie.

"Good morning to you, Sister Margaret," his diction altered by gingersnap remains. Each, without reflecting on it, used their formal titles in the presence of students. The hallway was full.

"I owe you a cup of coffee. Can I take care of that debt right now?" Maggie asked. "I get restless under any form of indebtedness, but especially when the IOU is with the male half of the universe." She was smiling.

Charles swallowed the last of the cookie. "Now is fine, Sister. I was just on my way to the cafeteria. I've been wondering whether I predicted the unpredictable episcopal mind correctly for once."

"You hit it right on the nose. Come, I'll tell you about it." They entered the cafeteria. Charles picked up a cold turkey sandwich from the rack, sniffed and rejected the steaming vegetable soup, and paid for the sandwich. Maggie signed for the coffee.

"Aren't you eating?" Charles asked as they sat down at a corner table.

"I'm on one of my sporadic fit-my-clothes diets, Charles." The trim Maggie was dressed in a lavender business suit, with the Franciscan Medallion of her Religious Order displayed on the silver chain about her neck. "I promised myself that all my clothes would fit comfortably by the end of this week. I splurged too much these last two weeks, can't pass up a french fry."

You're beautiful, Charles thought. Even the business suit can't camouflage that. "What's a diet?" he asked. "You get my vote for the best dressed nun around."

"Bless me, Father, it's my ego," Maggie intoned in the conspiratorial whisper of the confessional.

"Does the Order give you an extra allowance for clothes because of your job as president? Or shouldn't I ask?"

"It's OK to ask. The answer is no. There are no special perks in the job. The secret is in having a small selection of good quality pieces. Michelle Thorson at the Downtown Shoppe taught me how to shop. You buy a few good pieces that go well together and then you mix and match those pieces. I'm not rich enough to have a few good pieces in more than one size. Sooo, I'm into an occasional diet to insure the ongoing usability of what I do have."

"Ah, I see." Charles unwrapped his sandwich. "Tell me, how did it go with the excellent Frederick Patrick Sweeney?"

"Not well, Charles. I was very uncomfortable when I left the meeting." Maggie related the details of her meeting. "I question now whether it was prudent of me to comment on the role of women in the Church. However, the bishop opened the subject by referring to that article about me in yesterday's newspaper. He disagreed with the comments quoted from my speech so I had little choice but to respond. I tried to throw the orthodoxy issue back on your shoulders, and suggested the bishop meet with you and your staff. I knew you wouldn't mind, and might even enjoy the encounter. He sidestepped the invitation. It's the women's issue that sticks with me. I think I may have crossed a forbidden border. It's not that I said anything that isn't absolutely clear from the history of women in the Church. It's more like he didn't want the subject discussed. Oh, he was polite enough at the end of our meeting, but there was no mistaking a shift in his attitude, a coolness." Maggie plucked a loose thread from her sleeve. "I'm trying not to worry about it."

"I'm glad you tossed me the orthodoxy question, Maggie. After all, it is my bailiwick." Charles was thoughtful. "Did you ever wonder why he chose to talk to you on a matter of theology? Frederick Patrick knows that I chair the department. I suspect he was looking for a soft spot to get at the theology department without personal involvement. I doubt if he suspected your strength."

Inwardly Charles felt a growing anxiety for Maggie. If she had made an enemy of Frederick Patrick Sweeney, the bishop's next move would come on a playing field where Maggie would not even be permitted to compete. Nor would the bishop appear on the field. He would employ mercenaries. Charles finished his sandwich and they chatted while finishing their coffee. A bell announced the first exam of the afternoon.

"I hope our paths cross this summer, Maggie. If not, I hope you have a good one. Your plans include a vacation, I hope," Charles said as they rose to return their trays.

"I'll be spending a week with Sister Ann Johnson and Kate Corsalini out at the Corsalini farm. Maybe I'll see you there. I also plan to spend a few weeks with my family in Montana. I haven't seen them in ages. After a year of office, hallway, and classroom ceilings, I'm looking forward to the big sky country. How about you?"

"Tony and I plan a muskie fishing trip next week. Other than that I have no plans."

On leaving the cafeteria, Charles pulled a scribbled list from his pocket. He could not shake off his uneasy feeling.

CHAPTER FIFTEEN

The hostess showed Frederick Patrick Sweeney and Stanley Wright to their table amid the subdued elegance of the country club's dining room. They discussed the menus, gave their orders to the uniformed waitress and waited until she left the table.

"It's good of you to meet with me on such short notice, Stanley," Frederick Patrick began.

"Nonsense, Your Excellency. I'm always at your service. You mentioned that there might be some way in which I could be of help?" Wright's black hair, combed straight back glimmered under the lights. His impeccably dressed spare body housed a rich smooth voice. His mouth opened only slightly when he spoke and otherwise rested in a thin smile.

"Actually, there are two separate matters, Stanley. You can be of great help to me."

"I'll do what I can," Wright interposed quietly.

"I'm particularly pleased that you intend to serve on the Board of Trustees at Padua College. Do you know the president, Sister Margaret McDonough?"

"I've met her in social situations on a couple occasions, Your Excellency. I don't know her well."

"I have only recently had occasion to have an in-depth discussion with her, myself," Frederick Patrick continued, "and quite frankly, Stanley, I have doubts that she is the person for that job. I have particular concern for her judgment on what I call the Catholic identity of the college. So, I have need of your assistance, Stanley, to resolve my doubts. It is sad, but it is sometimes necessary to defend the Church against the poor judgment of its own solemnly professed. Stanley, as you know, there is no substitute for good judgment."

"I know exactly what you mean, Your Excellency. Judgment is important." In Stanley Wright's corporate world his subordinates damn well better have judgment, his judgment.

"The college isn't going to hell in a handbasket, Stanley. So there's no need for immediate and major surgery. As a member of the board you will be in a good spot to observe her judgment. And, of course, it would not be in the best interest of the Church if it appeared as if I were involved. I don't want any public fights that might indicate division within the Church. It just wouldn't look good."

"Rest assured, Your Excellency, that I will do my best for you. I am honored by your trust."

At that moment, the bishop's eye caught that of John Winslow Alexander who was seated in the midst of admiring women at a nearby table. The bishop returned a wave that Alexander took as an invitation. He walked over.

"Hello, John. Sit for a moment," the bishop said expansively.

"Your Excellency, Stanley." Alexander acknowledged each and sat down.

"School out, John?" Wright asked.

"Nearly, Stanley. This is exam week. Then it's over. How nice to see you both. Stanley and I go back a long way, Your Excellency. Back in our college days I dated Stanley's sister, Melanie. I was mad about her. It took me years to recover after she married another."

"But now you are an eminent professor with summers off. You have no family responsibilities and are surrounded, I see, by ladies. Seems idyllic to me," the bishop said sportively. "What are your summer plans, John?"

"I'm off to San Francisco next week, Your Excellency. Some research at the university there, and a bit of vacation."

"Stanley, we should be as lucky as this gentleman." The bishop turned again to Alexander. "Tell me, John, how are things at Padua College?"

Alexander tilted his head and sniffed. "I don't want to be uncharitable, Your Excellency. Suffice to say, the present administration seems to lack what I call vision."

The waitress appeared and set a stand next to the table. She left and returned with a tray. Alexander excused himself. "Time for you gentlemen to eat. It was good to see you both. *Bon Apetit.*"

The bishop and Wright ate slowly, engaged in small talk. Occasionally a woman or two would flutter up to the bishop and be dispatched with gracious and sufficient compliments. The dishes were removed and the two men faced each other over coffee.

"Bishop, you mentioned a second matter where I might be of help?"

"There is, Stanley. It is a delicate subject and..., well, I understand that you employ a priest who has left the ministry, one Anthony Corsalini?"

"That's correct, Your Excellency. Tony works for me now."

"It is a matter of concern to me, Stanley, because the presence of an unfrocked priest, if you will, in this city is an embarrassment to the Church. Frankly, it is an embarrassment to me personally. Here is a man who turned against his vocation, his personal call from God. I would not object so strongly, Stanley, if the man were unknown or kept himself outside the public limelight or if Mill Valley were a metropolis like Chicago. But, Father Anthony Corsalini was well known as a priest in this city. And, now, Mr. Anthony Corsalini is highlighted frequently in the press and other media for his involvement in one civic activity or another. Surely you can understand my position. It is a scandal to our good lay people. What must they think, these sheep of my flock. The Church should not be made to suffer this form of embarrassment."

Wright was defensive. "I should tell you, Your Excellency, that before I employed Tony I checked with your predecessor, Bishop Zimmerman. He told me personally that he had no objection whatsoever. In fact, he encouraged me to hire Tony. And, I must tell you, Tony Corsalini has been worth his weight in gold to me. I can give him any job in the company and have confidence that it will get done, and get done right."

"I'm aware that he is a talented person, Stanley. I'm also aware that he was a member of Bishop Zimmerman's inner circle, so to speak. I do think, however, that the bishop's encouragement to you on this matter was a singular lapse in judgment. Why, the very rescript from Rome that granted the man's dispensation requires that he locate himself at a reasonable distance from the area. I would think that the man himself would have had more sensitivity to the Church."

Both men reached for their coffee at the same time.

"I can remember an incident in Rome when I served as an editor of our seminary yearbook," the bishop continued. "One of my classmates wrote an article for the yearbook

106

which vividly depicted the poverty that was in and around Rome at the time. I will never forget the reaction of the rector of the college when he reviewed the article. He reminded us in no uncertain terms that one does not embarrass the Holy Father in his own diocese. Obviously, that article never saw print. The lesson has stayed with me. And I do not wish to be embarrassed in my own diocese."

Hesitancy showed in Wright's response. "I have just this week given Tony a highly commendatory job performance evaluation along with a substantial raise in pay, plus a bonus. It would be extremely awkward for me to go back to the office today and fire him."

"I don't know that I'm asking you to fire him, Stanley. Perhaps you can find some other way to lower his public profile. After you've reflected on my words, please take whatever action you feel is appropriate at a time that is suitable to you."

A waitress refilled their empty cups.

"And I thank you again, Stanley, for honoring the burdens I have placed on you today. Let us think of them as our secret and joint sacred trust." He paused and looked at Wright.

Wright nodded.

"I'm confident that God will be pleased and will reward you generously. And, if there is anything I can do in return, let me know."

"Now that you mention it, Your Excellency," Stanley took immediately to the opening, "there is something you can do for me. And here I must rely on your discretion. We both know that the plans for your new cathedral will soon be out for bids. Wright Construction personnel will carefully work on our bid for the job. I'll be frank, Your Excellency, I know my competitors. From my experience I know how their bids compare to mine. I know that my bid will be within $100,000 of the lowest bid. On a project of this size that's peanuts. You're a builder, Your Excellency. You understand contractors. They can underbid by that much money and make it up in a dozen ways: shoddy materials, shave a little here and there by putting pressure on subcontractors, all that sort of thing. Wright Construction prides itself on being a Catholic company." He waited.

"I think I understand, Stanley."

"Thank you, Your Excellency." Wright signed for the check and the two men walked out of the club together. In the parking lot they encountered Nick Deutschman, dressed for golf.

"Now here's the diocesan expert in education, Stanley. Did you know that Monsignor Deutschman has a doctorate in educational administration?"

"No, I didn't, Your Excellency. I only know how smart he is at construction contracts."

Deutschman absorbed the praise, acknowledged the bishop with a bow, and hit Wright in the shoulder with a manly jab. "Can I interest you gentlemen in eighteen holes?"

"Not for me, Monsignor," Wright responded. "I don't work as fast as you. My desk is still loaded."

"Nor me, Nicholas. One round per year is my limit," the bishop said jovially. "I appoint you to uphold the honor of the priesthood on the links."

Deutschman headed for the pro shop. Wright walked the bishop to his car. He opened the door. "I hear that you're leaving for Rome shortly."

"That's right, Stanley. I will touch base with several Curia members. After that, I have two weeks of vacation planned. I will visit galleries in Florence, Venice and Zurich. You know my weakness there."

"It's about time that you took a vacation, Your Excellency. You deserve it. You've been working too hard." Wright pulled a stuffed white envelope from his breast pocket. "I'd like you to have this as a token of my respect and appreciation for all you do for the Church here in Mill Valley." He handed the envelope to the bishop. "Now, I want it understood that this is a personal gift to you. Cash can be a great convenience in some of those European shops and galleries."

He paused. "I will not take full credit for this gift. Part of it will come from Uncle Sam if you will be kind enough to send me a receipt on your diocesan stationery for a charitable donation. The Internal Revenue Service makes such a fuss if one's documentation isn't in order. My main concern, however, is that you have a successful and delightful trip."

"Thank you, Stanley. You are indeed generous. I can't tell you how strengthening it is for me to know that I have the support of a man of your caliber."

Upon his return to the chancery, Frederick Patrick emptied the envelope and penned a note of thanks to Wright that included a receipt for a $10,000 charitable donation.

CHAPTER SIXTEEN

Charles turned and reached for the brown paper sack in the back window, opened it, and looked in. "I knew it. Gertrude wrapped the rest of the cake for us." He pulled out an individually wrapped section, leisurely undid the plastic wrap, and started munching. "Austrian poppy seed cake, chocolate filling, I'm in heaven. Too bad you're driving, buddy. I'd give you some but you'd get your fingers all gooey."

Charles and Tony Corsalini had stopped at Frosty's country rectory for lunch on their way north. Gertrude, God bless Gertrude, had filled them with food. Frosty had filled them with fishing advice. "Ya, boys, you just remember the three S's. You want to catch muskies in June? It's shallow water, small bait and slow retrieve. You got that, boys?"

"How can you eat more?" Tony asked. "I'm so stuffed I could pop."

Charles ignored the question and continued his fascination with the cake. "Wasn't that meal something else? The incomparable Gertrude and her German cuisine. It's a good thing Frosty's got that huge frame to hang all those calories on."

The pickup truck belonged to Tony and Kate Corsalini. Charles and Tony jointly owned the fifteen-foot Alumacraft they were towing and the forty-horse outboard motor on the boat's stern. They rode in comfortable silence as they watched landscape change gradually from farmscape to lakescape, and from the uncrowded newly greening ash and maple woods to the dense green of fir, spruce and pine.

They were classmates, friends since their first meeting in the seminary. And it wasn't just that silence was comfortable for them. It was also that now was not the time for words. It was a time for letting go, for discharging the winter's accumulation of stress, the cabin fever that accompanies life in the snowbound season. It was time for a psychic massage, the kind that nature kneads when you open your senses to it.

"That monster muskie is awake and hungry," Tony had enthused when he picked Charles up that morning. "You realize, Carlos, this is trip number thirteen for us? That means *bona fortuna,* eh? We're gonna catch thirteen muskies this year. I can feel it in my bones."

Charles returned the brown sack to the back window ledge.

Tony, thin, muscular and athletic, drove with eyes fixed to the road. His slightly creased brow rose into a half saucer of baldness.

Charles looked at Tony and recalled how crinkled that brow had been when Tony had revealed that he was leaving the active ministry to get married. The wrinkles had disappeared with Charles's ready acceptance. That was nearly five years ago. They had talked about celibacy. Both had accepted it as a requirement for ordination, taking as timeless wisdom that it was required to keep the heart and mind centered on the work of the Lord. But, their experience did not agree with timeless wisdom. Each man knew many priestly celibates for whom a half dozen phone calls constituted an exhausting day of work. And both knew dozens of married people who were more productive on their lunch hour than those celibates were in a day, and who had a deeper spirituality, a more singular focus on God.

Charles closed his eyes for a minute. He wondered again how Tony, whose usual burbling tongue had always been mute around women, had ever articulated his feelings to Kate. He pictured Tony's face at the wedding, joy radiating, and asked himself, what does that kind of joy feel like?

Tony, alert to the road even while the back channels of his mind filled with memories, reviewed names he had settled on his friend over the years: *Charles the Silent* for his ability to be at ease with silence; *Charles the Feckless* for the noise he made while fishing; *Charles the Undernourished* for the pockets full of gingersnaps. He broke the silence. "You know something, Carlos, you read like the boy scout's code: trustworthy, loyal, helpful, friendly, courteous, kind, cheerful, obedient, thrifty, brave, clean and reverent." He reeled the words out in the same monotonous rhythm he had as a thirteen-year-old scout. "What you need is a little excitement in your life. I can't believe I didn't see this long ago. I finally got you figured out. I'll bet your DNA is imprinted with the scout's code."

"A prophet you're not," Charles replied, laughing. "For your information, I flunked both obedient and reverent, got a D in thrifty, and I'm still waiting for a grade in brave. You'd think, after all these years you'd know me better, if you had the least bit of talent for that sort of thing."

"I read you like a book," Tony retorted.

"You're still reading comic books."

The two men fell silent again. Tony recalled Charles's struggle whether to stay or leave the ministry when they were together in the chancery, when the clear ideal they had shared of priesthood as selfless service had collided with the reality of the clerical world. What keeps him there? he wondered. Maybe this year they would get to that. A related thought prompted Tony to again break the silence. "Did you see the editorial Sam Duffy put in the weekly wipe last week?"

"In the what?"

"The diocesan newspaper."

"Oh, *The Catholic Courier*. No, I didn't read the editorial. I only read the section that lists priests' transfers. Otherwise, there's just too much applause for the clergy: pictures of bishops on every page, father's 25th, 30th, 40th, 50th anniversary. What's with the editorial?"

"It's about why priests leave the ministry."

"Why do they leave?"

"Sam gives a string of reasons: the guys are narcissistic and sophomorically immature, and/or sex crazed. They are delinquent in their spiritual exercises, or they've lost their faith."

"You're putting me on." Charles looked incredulous.

"Not even *un piccolo*," Tony continued, "and Sam denies that he's violating charity. He says the Holy Spirit doesn't give a vocation only to withdraw it later."

"That surprises me. As if it's unthinkable that the Holy Spirit, through those tens of thousands of departures, might be trying to get a message through to the institution. Hey, slow down, we're here." Charles pointed at the sign on which a huge muskie, curved like a quarter moon, jumped clear of the water, hooks from a bright purple bait stuck in its jaw.

Tony turned the vehicle into the Muskie Cove Resort at the south end of the 2,700-acre North Dog Lake, home to elusive muskies. Tony honked the horn at the resort's owner, Genial George, who was busy repairing a scratched and dented Johnson outboard. George walked over.

"You fellas is just in time." George shook their hands. "It's been dead, dead, dead, until yesterday. Yesterday, one fella got three nice ones, another guy got two of 'em. Everybody was gettin hits. So, you fellas get unpacked and get out there. Cabin six is already set up and waitin on ya, just like always." He ambled back to his work.

Gear and grub stored, they readied the boat, stowing rods, net and tackle boxes in place. Standing on the dock, Tony leaned over the edge to observe a school of bait minnows. Charles, from behind, jabbed his hip into Tony's rear. Arms flailing, Tony hit the water on his belly, went under, and came up dripping.

"You here to swim or fish?" Charles asked with fake innocence.

"God, this is cold. Help me out of here." Tony, shivering, held out his hand for a helping hoist. Charles grasped the hand only to be pulled off balance and duplicate Tony's belly flop. As Charles surfaced, sputtering, Tony said casually, "I'm done swimming now, shall we fish?"

They changed into dry clothes and in short order were on the lake. "Five bucks says I get the first fish?" Tony asked.

"Legal or any size?"

"Legal."

"Hooked, or in the boat?"

"In the boat."

"It's unethical to bet on a sure thing, but as long as you've been warned, it's a bet."

They fished crank bait style, standing, one in the bow and one in the stern, as they cast the artificial bait as far as they could and cranked it back.

115

Charles saw the fish when it was about fifteen feet from the boat. Coming on fast it followed his bait, then stopped. Charles, rod tip in the water, made the mandatory figure eight with his bait. The muskie backed off and disappeared.
"Son of a gun, I thought that fiver was in my pocket. Just had a follow," he said to Tony. He felt his heart pound. "Lets you know there are fish on the prowl." A half hour later, Tony pocketed the five dollars after landing a thirty-six-inch fish as they worked the edge of the Pork Chop bar. The battle lasted less than five minutes. A few zigs and zags and the fish was in the boat.

"Just my luck," Charles said, "you meeting up with the laziest and dumbest muskie in the lake when there's money riding on it."

Tony extracted the hook and pulled the fish from the net.

Charles pulled a camera from his tackle box. "We've still got enought light."

"Hold it," Tony ordered. He switched the fish to his other hand. "My right profile is my best side."

"A best side, you haven't. Smile."

"I'm smiling."

"Not you, the fish."

They released the fish. Dusk was falling fast as they turned the boat toward the cabin.

Their routine had become ritual over the years. They unstuffed grocery sacks and unpacked the cooler. Tony held aloft two bottles he had just unpacked. "Look at that clear red color. Corsalini's own and latest homemade Dago Red. Aged in the bottle at least thirty days. You're gonna like this stuff, Carlos."

116

They would dine out on the first night. After that, Mueller was cook, Corsalini was bartender and dishwasher. Lettered under a layer of shellac the board nailed over the stove read: *Fortune's Favorites Fish From a Full Belly.* They took it seriously. That first night they dined at the Holoway Bar, four miles down the road. "Walleye pike don't come no better than at the Holoway," Genial George had told them on their first trip. They agreed. Home from the Holoway they cracked windows to let in the crisp night air and the deep northwoods silence, broken randomly by the stirring cry of the loon. They turned in at 9:30 p.m.

The next morning, Charles sat on a boat-sized boulder and watched the dawn come on. The osprey caught his eye. He saw it hover and then dive, braking as it touched the water, coming aloft again with a squirming fish held fast in its talons. The osprey gained altitude and winged toward its nest. It was a squatter on the top of an electric power pole. Then Charles saw the eagle. It dove on the osprey. The osprey dropped its prey in a protective maneuver and the eagle snatched the fish. A sense of wonder and privilege from witnessing the event filled Charles for a time.

Which are you, Mueller, he thought, the osprey or the eagle? Not the eagle! You don't know when to take or how to take. Not the osprey, either! That osprey didn't give up the fish voluntarily. You're the fish, Mueller. That fish didn't have any choice about ending up as bird food. You don't have a choice either. You're a compulsive giver, Mueller. You let people suck you dry and you call it virtue. Tony had a point with his boy scout code remark. You've got to learn when it's right to take. If you don't know where to start, start here with nature. Nothing to fear here, Mueller. Nature gives itself gently on sunny stormless days; sunburn maybe, a mosquito bite maybe, but nothing you can't take. Practice taking, Mueller. Let the smells of fish and flowers in. Let the sound

117

of the loon stir you. Let the sights of land and water move you. That'll teach you about God, Mueller, and how to take what he gives.

Back in the cabin Charles poured water into the percolator, sniffed the freshly opened coffee, and spooned it into its holder. He knew its aroma would be alarm clock enough for Corsalini. "Come on Lazarus, it's time to fish," he called when he heard the first footstep. "We've got perfect weather."

Tony came straggling out of his bedroom and went into the bathroom. Minutes later he emerged, and stood proudly in his shorts. "Escremento magnifico!" he exclaimed.

"Enjoy the anal stage while you've still got it, Tony. Now, how about breakfast?"

* * * *

The boat drifted a course laid out by the wind. Tony used the motor only to drive them back for a drift parallel to their last one. They fished the eight- to twelve-foot depths. It was midafternoon when Mueller yelled, "I've got one." His rod bent. He cranked, but the line stayed stationary, neither coming in or going out. "Might be a snag." Then it moved. The reel whirred as twenty yards of line spun out against the drag. "Whoa, it's like I don't have the drag set." It was set, he knew, balanced heavily against the 20-pound test line. "This could be the big one." Twenty minutes later the contest stood as it had started. Charles got some line back onto the reel; the fish took it back. "I'm getting pooped." At one point only several revolutions of line remained on the reel. Another twenty minutes went by. "We're getting somewhere." The fish came directly at the boat and under it. Charles struggled to keep the line clear of the motor. "Should see it pretty soon." Then it was there, near the surface, thirty feet away. "Wow, look at that baby. Gotta be forty, maybe fifty pounds." The fish dove and then came up and rode the surface, its head clear of the water, shaking from side to side. Then it submerged again.

"You've got it. It's tired. Keep it coming," Corsalini encouraged as he grabbed the net.

Then it happened. The big fish surfaced and flipped its huge head one more time. The bait popped out as if some invisible arms had worked the Heimlich maneuver and disgorged the throat-choking chunk. Charles and Tony watched as the fish lazed deeper and out of sight. Neither spoke. Disappointment hung like black clouds on pine tops. Charles searched his pockets for a gingersnap.

So complete had been their absorption with the fish, they hadn't noticed the weather change. The wind was rapidly gaining strength. There were black clouds coming onto the lake over the pines. The lake grew more perturbed and sprayed white windward from a hundred yards off the lee shore. "Time to get out of here," Corsalini exclaimed as he started the motor. The rain started as they docked.

Rain didn't keep them off the lake, though it continued through the next two days. Wind kept them off, causing chop so rough they were unable to stand in the boat. They watched from the cabin as the storm ran its course. Both men were readers and had come prepared with stacks of mysteries. They read. They talked. The talk started on the edges and worked its way to the center. Corsalini was the first to reach the center. It was cocktail hour, the second day of rain.

"Did you ever notice that laymen make better drinks than the clergy?" Tony asked as he took the first sip of his Rob Roy. "That's my brilliant lead-in. Speaking of the clergy, you have told me more than once that the smell is destroying your nose. How come you're still hanging in there?"

Charles tasted his Manhattan and thought for a moment. "It's the blasphemy," he said, eyes twinkling.

"The what?"

"The blasphemy."

"I imagine you're gonna tell me what you mean?" Tony asked.

"Just that. I'm hanging in there because someone ought to, just to point out the blasphemy."

"Is this twenty questions or something? What blasphemy?"

"Clerical blasphemy. You know. Where they turn God into their personal servant. You know what I'm talking about, Tony. You've seen it." Charles aimed a stilted oratorical voice at the rafters. "Almighty Gawd demands of you, his faithful people, under pain of sin and hellfire, that you squat when we say squat. And by the way, the rectory needs refurbishing. And don't forget that the clerical discount is as obligatory as a Holy Day. And don't forget that we can make it a sin to eat oatmeal. And above all, don't forget that Almighty Gawd has put us up here and you lay folks down there, so shut up and listen."

"I get the point," Tony said. "Did I ever show you the rescript of my dispensation from Rome? I was *reduced* to the lay state. Honest to God, exact words, *reductionem ad statum laicalem.* That says it all. Lay folks stand on the bottom rung. If they look up, they can see clerical buttocks. I'm still a priest, but now I'm a layman. That's what you're talking about?"

"Right." Charles took another sip and continued. "And those self-serving exaggerations only get worse the further you look up the clerical ladder. At times I think our church leaders differ little from a cadre of ancient Mayan priests, shouting their ritual babble as they flay another victim. What suffers is the priesthood and those of us in it. Those clerical extortions cling to it like parasites and leech away its credibility."

Tony thought for a minute. "Sounds to me more like reasons to leave than stay. Back to my first question, why do you stay?"

Charles spoke slowly and thoughtfully. "I don't know that I will stay. Right now I'm hanging in there because I believe in the priesthood. I believe that a man can be a priest to the people, even carrying all the clerical debris. There are a lot of men who are priests and the cleric doesn't show through."

He stopped and stared at his drink. Then quietly, "I am torn though, pulled hard toward marriage. I don't believe celibacy necessarily goes with the job, but the hierarchy won't break that connection in my time." Charles pictured Maggie and felt the force of her pull.

He looked at Tony. "I'm trying to work it through, but I'm not there yet. I'll hang in there until I work it out more clearly. Besides, I learned on the farm that bad smells come with most jobs. What about you, now? You seem to have it all together. I've never seen you more consistently cheerful."

"Carlos, you are looking at one happy man. I never imagined that life could be this good. As you were talking I realized that our creeds are the same. We found the same

121

God, the God of care and affirmation. The only difference; I found Him in Kate's eyes, the way she looks back at me. I know that I didn't understand love or God until Kate. I believe, as you preach, that human experience is God's native tongue. It's the only way he speaks to us. My experience differs from yours, but we both came to the same God."

"I saw Kate at the hospital a week ago. She looks as happy as you do." Charles paused at the sound of a motor on the lake. "We'll be fishing tomorrow," he said.

The next morning they found the lake muddied by the storm. They fished without success.

"Hey, Carlos, get a fish will you. You're supposed to be the four-leaf clover around here."

"Four-leaf clover? I thought you Italians had all the charms and spells. Touch your belt buckle when you see a priest so you don't go impotent, that sort of thing. You got any spells for fishing?"

"*Niente.*"

"Say, Tony, how much Italian do you know?"

"I don't know. Maybe twenty, thirty words. Why?"

"You might like to know, Carlos is Spanish. Good thing you don't need the Italian for your job. Which reminds me, you haven't told me about your job. How's it going?"

"Great! I now hold the titles of office manager and chief expediter at Wright Construction. Translated, I'm chief gopher. But I enjoy it. There's a crisis every day. Work is never dull and I seem to get along well with the boss."

"What's Stanley like as a boss?"

"On the outside Stanley is Mr. Congeniality, easygoing and homespun. On the inside the wheels never stop. He's a calculating, shrewd businessman. On several occasions I've seen that congeniality quickly turn to ice toward one or another of the employees. But, like I said, I seem to get along well with him. Maybe it's my clerical training. I'm paid well by Mill Valley standards. In fact, I just got a heck of a raise in pay and a big bonus. So, the job is stable.

My only problem is that I'm putting in sixty plus hours each week and that's a bit much for a family man. And the job keeps growing. I've told Stanley that I'll need an assistant soon. Last week he brought one of his nephews into the firm. I don't know him well yet, but he strikes me as Stanley's clone. I'm supposed to train him. Maybe I can get some help from him."

"Do you miss the priesthood?" Charles asked.

"Not for a second. Hey, look at the time. Let's get ashore. Are you going to feed us tonight, or do I have to make myself something tasty for a change."

"You? Cook? Remember that one time I let you cook? You made what you called hunter's stew. George's dog wouldn't eat it. Besides, I promised Kate not to let you near the stove. She's afraid you'd burn yourself."

CHAPTER SEVENTEEN

Frederick Patrick Sweeney sat in the barren conference room at the congregation's headquarters on the Piazza Pio XII. His straight-backed wooden chair with its slight rise of tapestry cushion was uncomfortable. The faint musty smell of old stone filled the room. Early for his appointment with the cardinal prefect, he studied the room's sole ornamentation, the cardinal prefect's space-dominating portrait. Alberto della Tevere, cardinal prefect of the Sacred Congregation for Bishops, tall by Italian standards, and slender, stood fully vested from the skullcap zuchetto to the flowing train of the cappa magna. The bishop's mind folded back to the *disputatio*, an exercise in theological reasoning, at the Gregorian University. Both men had been observers. Frederick Patrick, the seminarian, had introduced himself to della Tevere, then a young monsignor in the office of the secretary of state. Taken by the student's American-style exuberance della Tevere had invited Frederick Patrick to visit him at the secretary of state offices in the Vatican Palace. A first visit led to others and to a matchless introduction to Vatican ways.

"To understand the Vatican, you must understand patronage, Federico," he had said. "It is the convention that underlies all human transactions here." Frederick Patrick could recall the conversation almost verbatim. "Federico, the *patron* does not say, 'I am *patron*'. It is rather the countless minute and major, silent and overt, requests for favor and favors rendered that mark him as *patron*. It is no sin to be ambitious, Federico. But, if you choose ambition you must learn this skill. Forget about democracy, Federico. Forget about merit. For the cleric who wants to advance his career, whom you know and who likes you is of the utmost importance.

Of course, one must have the ability, the intelligence, but most important, one must be skilled in the courtly art of patronage. One must learn, Federico, how to place an ever so slight sense of indebtedness on the *patron* in such a way that the *patron* takes delight in both the manipulation and the indebtedness. It is a bribery, Federico, that leaves no evidence of the bribe. The *patron* is left with the feeling of power. Power, I think, Federico, is the experience that rushes blood to the clerical groin." As it turned out, della Tevere himself had become Frederick Patrick's *patron*.

"Caro Federico, it is so good to see you." The cardinal swept into the room jolting Frederick Patrick abruptly from his thought and to his feet.

"Buon giorno, Eminenza." Frederick Patrick barely got the words out before the cardinal embraced him.

"You are well, yes, Federico?" Della Tevere stepped back and looked closely into Frederick Patrick's face through round metal-rimmed glasses. "Yes, I can see that you are marvelously healthy. That is good. Come and sit over here. We will speak the English, yes?" He led the way to chairs at the table's end and took a seat. "Before anything else, Federico, I must thank you for the delightful assortment of your cheeses. They are wonderful." He brought joined finger tips to pursed lips in his country's gesture of gastric delight. "And they were all made in your diocese of Mill Valley?"

"It is nothing, Your Eminence. I am pleased that you enjoy it. My diocese lies in dairy country, and produces many varieties. We're peopled with emigrants and their descendents from all European countries and each nationality has adapted its own cheese-making heritage."

126

"You must be careful or I will be pestering you for more. I have such a weakness for the cheese. But, tell me now, Federico, are you enjoying your visit to Rome?" The cardinal's slender hands and long fingers choreographed his words.

"I am, Your Eminence. Rome is as captivating as ever. I must confess, however, that my experience here in the Vatican is different from my student days. As a student I felt secure here. Today, even as a bishop I feel somehow intimidated and insecure. There is a sense of mystery here that I never experienced before, the sense that so much is hidden from view."

"*Non ha paura,* Federico," the cardinal replied, leaning forward and patting the bishop's forearm.

Non ha paura, do not be afraid. Frederick Patrick remembered hearing those same words from the doctor just as the anesthetic had begun its work. He had broken his big toe playing handball, the toe corkscrewed from abrupt contact with the court wall. "*Non ha paura,*" the doctor had said as the anesthetic forced Frederick Patrick into unconsciousness. That uneasy experience from his student days was not all that dissimilar from his present discomfort. Not a comforting comparison either, as the toe no longer flexed.

"It is true, Federico," della Tevere continued. "There is secrecy here. And there is fear, too much fear. It seems that fear is the air we breathe. I see how it withers the eyes and features of so many. Fear and secrecy are indeed companions. But you must not have fear. I am on many councils here. There is not much that is secret from me." The cardinal raised his long fingers to the sides of his head. "You will use my eyes and *non ha paura. Va bene?*"

"You are always so kind, Your Eminence. Thank you. I am always comfortable with you."

"It is nothing, Federico. We are friends, are we not? And perhaps someday together we can change things. This secrecy is a great sadness to me. If there is any place in the world where truth should be transparent, it is in the Vatican. If there is any place where discussion and disagreement should be open and fearless, it is here. If there has ever been a message that is the, how you say it, opposite to secrecy and fear, it is the gospel. But, I talk too much." He leaned back in the chair. "Now tell me about you and your Mill Valley."

"You must visit us someday, Your Eminence. The countryside is very beautiful." The bishop talked about his diocese, his educational programs, his building programs, and his concerns.

The cardinal listened intently, eyes focused on Frederick Patrick's face. What a marvelous talent this young man has, he thought, how politically effective. He will need a shift in direction, but his loyalty and obedience will provide for that. Aloud, "I am impressed indeed, Federico. Your talents are evident. We must only be sure to remember that, of all men, we bishops especially must be champions of individual rights. But, we will talk more of this."

Frederick Patrick showed his puzzlement, but the cardinal went on. "Federico, I must get back to my other appointments. Tomorrow evening I am dining with several close friends at the Trattoria San Stefano. It is near the Piazza Navona. I would be pleased to have you join us and meet these friends. They will like you. You will like them. We have a private room so give my name to the host. It is at eight o'clock. You will come, yes?"

"I would be delighted and honored, Your Eminence."

The cardinal walked the bishop to the portico. "Addio, Federico, until tomorrow."

"Thank you, Your Eminence." Frederick Patrick walked out of the building onto the sun-baked Piazza and turned into the Via della Conciliazione. His home for the stay, the Hotel Columbus, was only a short distance down the block.

I don't miss this heat, he thought, remembering how his student summers had been spent traveling or at a Villa in the cool Alban Hills below Castel Gandolfo. The hotel, impressive with its Renaissance facade, was even more impressive in its fifteenth-century interior. He welcomed the cool of the interior courtyard as he admired its colonnaded arches. He had chosen the hotel with care. It does not rank among the deluxe hotels for comfort or price, he thought, but it is comfortable. And I will not be seen as farting higher than my own arse, remembering the advice of his grandfather, and his own regard for Vatican discernment.

It is impossible to measure time in the eternal city. The thought occurred to Frederick Patrick, not as the product of philosophical musing, but as his taxi careened down the Corso Vittorio Emanuele and then pointed north toward the Piazza Navona. The taxi missed collision with other cars, scooters and pedestrians by scant inches, the driver refusing to insult his machine by slowing down. Frederick Patrick had given himself ample time, not wanting to be late. Even so, he had not expected to be deposited in the piazza a full thirty minutes early.

Shade was filling the western side of the piazza's baroque perimeter, which retains the original dimensions of Domitian's stadium. The bishop closed his eyes and pictured a Ben Hur style chariot race. My favorite piazza; I'll walk the center, he thought. He recalled the Christmas displays and vendors of Christmas wares from his student days. Walking up the rectangle, he skirted the Fountain of the Moor, water belching from conch horns blown by kneeling figures. He stopped to examine the facade and bell towers of the Church of St. Agnes in Agony, its stained windows dark from the wrong side of the sun. I said Mass there, how many times? He visualized the interior Greek cross design. Moving on, he stopped and sat on a marble bench next to the piazza's highlight, Bernini's Fountain of the Rivers. Rather than evoke a sense of peace and calm, the falling waters triggered a sense of unease. He suppressed the feeling. This is not like me. I will be as carefree as a student. I'm going to a party. He rose from the bench and walked the remainder of the piazza to the restaurant and to a room filled with laughter.

"I see that Ludwig has found himself another witch to burn."

"Who is it this time?"

"Katerina Klug. Degree in theology from Tubingen, she teaches at Upsala."

"What is he using to fuel the fire?"

"She is trying to make the Blessed Virgin human."

"Ah, I think Ludwig would prefer to have all women silent, pregnant and in the kitchen, as the saying goes, but especially silent."

Frederick Patrick knew they were talking about Cardinal Ludwig Schwindel, prefect of the Sacred Congregation for the Doctrine of the Faith, formerly the Holy Office, and at its origin, the Sacred Congregation of the Holy Inquisition. The conversation brought back his sense of unease. How can they talk so flippantly about a person of such eminence?

They were seven in all, including della Tevere. The cardinal had introduced him to each man individually. He had heard of them all, all prelates, and all but for della Tevere in the second tier of Vatican prominence. They were powerful in the way that executors wield power, like sergeants in the military. Their job was to execute the papal will, but he knew that each had the skill to put his individual turn on that will.

"Claude, now that more than a year has passed since Archbishop Romero was killed in El Salvador, do we have any clear picture of how the Church came off?" The question came from the Scot, Archbishop Timothy Burns, a short, stocky, terrier-faced man. It was addressed to Claude Dupuis, tall, with a thick black mustache, hair in disarray, and doleful eyes that complemented slightly drooping jowls. Claude was an archbishop in the office of the secretary of state.

"We scourge the poor for our rich masters, Tim. Some of us at State got beat up badly on this and lost. As you know, Romero changed over the years from an old-line conservative into a champion for his poor. The new government in El Salvador consists of Catholic intellectuals and the military. It was and is the rich against the poor. The Americans saw a threat to their pocketbook from the rise of the poor, and, against Romero's advice, supplied arms and support to the ruling junta.

The junta began the killings. Thousands are already dead. They paint it as an anti-communist action for the obvious political benefit. The Church, officially, has sided with the rich on the basis of opposing liberation theology, which champions the cause of the poor. Romero was murdered because he defended the poor. To sum it up, the killings go on, supported by the Americans, perpetrated by and for rich Catholics--who are blessed by the Church." The Frenchman's eyes grew more doleful at the thought.

131

"So once again the Church comes across as the oppressor of the poor and the guardian of entrenched wealth." The Scot's face reddened.

The tall, smooth-skinned Nigerian, Francis, from the Congregation for the Causes of Saints, broke in, his high forehead grained with a temporary crease. "And don't expect Romero to get the martyr's crown. A publicity campaign, complete with unofficial information leaks, is putting out the word that he was killed by his own supporters in order to claim political capital. So no martyr, that one, at least in this current dispensation."

Della Tevere raised his hands to shoulder height, palms outward, his slender fingers spread in a calming gesture. "There is much that we will do. For now we must be patient and we must plan." His calm was contagious.

Frederick Patrick sat stiffly. His unease counterpointed della Tevere's calm. What sort of group is this? he wondered.

The group muted as waiters appeared to clear the table of assorted appetizer dishes, empty now of their mozzarella marinara, spiedini, melon and prosciutto. The waiters vanished only to reappear with trays of entrees. Frederick Patrick ordinarily loved the smells that emanate from Italian cuisine. They were his favorite. Now, his appetite deserted him as the cannelloni was placed before him.

"Francis, I hear you are about to canonize Pope Alexander VI over at Causes of Saints. The word I have is that if Pius X could make it, Alexander should be given a shot at it. I hear also that you have credited him a miracle for each of his bastards and one for the hanging of Savonarola." The doleful eyes had turned bright with mirth in the placid countenance of the Frenchman.

"Actually, Claude, we've put a priority on Pius X's horse. We think it will speak to horse lovers everywhere. Alexander VI has been pushed back." The Nigerian's quick response parried the verbal jab with like good nature.

"Federico, you will forgive these jesters, yes?" Della Tevere's keen eyes had missed nothing. His raised fingers danced with the laughter. "Let me explain what we are about. It is not fair that you are in the dark. I have told these illustrious clowns all about you. They take my word. We would be pleased to have you join us if you choose to do so. We are men committed to the Church who believe certain changes are required in the institution. We call ourselves *Il Cero*, the candle. We have decided to be one candle and see if the Lord will let us give our light to His Church. The Lord Jesus is not always visible in certain institutional structures and processes. In fact, at times it seems that He is positively excluded. We seek to change that. We do what we can now, but it is little. Our hope is with the future. It is not such a bad thing, yes?"

"I am honored, Your Eminence." Frederick Patrick groped for the politically correct.

"With these friends, Federico, I am Alberto. When we are together we use only our first names. Try it. The strangeness will disappear quickly."

"Thank you, Your, Em...er, Alberto. I confess to some apprehension in such august company. But with such fellowship, that too shall pass, I'm sure." He did not sound sure.

"August? There is no one here by that name, Frederick." The Scot raised the temperature to a still more comfortable level.

By dinner's end Frederick Patrick felt comfortable with the company, if not the banter. Was it subversive or merely irreverent? If subversive, to whom could he report? The power of this group could easily counter any tattling, and make him look the fool in the process. Subversive? Ridiculous. Alberto is a prince of the Church. He knows the Church as the pope knows the Church. And I have known him all these years. He would never betray the Church. Trust Alberto and, as an afterthought, these good men. They are offering you a ladder to success. Take it. To refuse will hang the millstone of Mill Valley around your neck for life. Play the game, at least until you know more.

Della Tevere accompanied Frederick Patrick out of the restaurant. "Let us taxi together, Federico. I have more to tell you." He signaled for a taxi and gave instructions to the driver.

"Federico, you understand that the laughter and seeming irreverence of these men relieves the stress. These are dedicated men."

"I have no doubt of that, your, ah, Alberto."

"It is our purpose, as I said, to examine all the structures of the institution to be sure they speak Christ to the world. We are not revolutionaries who want only to tear down, if that is a concern to you?"

"I have known and trusted you all these years, Alberto."

"Would you like to join us then? You are politically very effective. We will have much need of that talent. We think also that you can become our expert, our *perito*, in managing conflict. In change there is always much conflict."

"I am deeply honored. Barring only the claims of conscience I am yours to command."

"No commands, Federico. Invite is the word. Like with our Lord Jesus, who invites us to follow." They rode in silence for a time. "Federico, there is no hurry for you to decide. *Il Cero* will meet for a week's retreat two years from now to review our research. In the meantime, there is much to get ready. Each man has been assigned an area of preparation. Because you are not in Rome yet you will not be burdened for now. And, from time to time I will send materials to you for your study. They will give you the direction of our research. If you decide as you read the materials that you cannot accept our invitation, that is your right. We will still be friends, yes?" The taxi pulled up to the Columbus Hotel. "Now, you will be in Rome for several more days, yes?"

"Yes, I will take the train to Florence next Monday."

"Good. I would be pleased if you would dine with me again this Friday evening. You will meet a different group of Vatican officials. You will come, yes? After that I promise to leave you alone for the rest of your stay."

"I am honored again, Your Em...Alberto. Of course I will come."

"Good. This time we will dine at my apartment at eight o'clock. This card has my address. In the meantime, remember Federico, *non ha paura*." There was warmth in his voice and smile. "Oh yes, one more thing, Federico, we try to be discrete about *Il Cero*, at least for now. You understand, of course?"

Walking into the hotel, Frederick Patrick's head was echoing the cardinal's words, "Because you are not in Rome yet." Not in Rome--yet? He was thrilled by the idea--and chilled.

A night's rest did not dispel his sense of unease. What am I nervous about? he wondered. That's not me. "Be discrete," he had said. Why discrete? Secret? Secret and subversive? No, never! Thank God for the diversions.

The diversions entailed taking his assistant, Father John Curley, along many of the paths he had followed with his camerata in his student days. He forced his thoughts in that direction. He remembered the men of his cameratas. Annually the student body had been divided into cameratas, groups of seven or eight, one or two from each class year. The camerata was obliged once per week to exit the college and visit some historic or cultural site. It was through those walks that Frederick Patrick had imbibed the myths and legends, truths and history of Rome. Thank God we've finished the 110 acres inside the Vatican he thought. I will be more at ease, hopefully, outside the Vatican walls in Rome itself.

Curley was enthusiastic. In many ways at thirty-two he was still the boy, the good boy, he had been at eight. He had goggled at the majesty of St. Peter's Basilica and Michaelangelo's ceiling in the Sistine Chapel. Alone, he had plowed through the Vatican museum when the bishop was otherwise occupied.

For two days Frederick Patrick almost lost himself in the role of tour guide in Rome. Almost.

The apartment of Cardinal Alberto della Tevere was located on a side street near the Piazza Risorgimento, a scant quarter mile from his office. It was spacious, but monastic in its appointments. Except for its size it gave little sign of the power and prestige of its occupant.

All of the guests had arrived by 8:30. The last to arrive was the Vatican's gray eminence, Cardinal Ludwig Schwindel. Frederick Patrick knew the curial roster and he recognized everyone. These were first-tier men, after the pope, notches above the Il Cero group, della Tevere the only exception. Eight of the assembly were cardinals. Frederick Patrick was unaccustomed to feeling himself a country bumpkin, but feel it he did. The feeling was countered in part by the warm introduction given him by della Tevere to each man individually. "Federico is my dear friend. We have been friends for years..."

Frederick Patrick found himself placed near the center of the long, narrow dining table. Directly across the table was the cardinal prefect of the Congregation for the Doctrine of the Faith. The austere German held and deserved the reputation of Vatican hatchet man. If the Vatican possessed a second Richelieu, it was this man. For a brief moment Frederick Patrick felt the tug of panic. *Non ha paura*, he repeated inwardly and then gave himself to the thrill of being in such company as he listened to Schwindel display an easy cordiality. While waiters scurried about placing and removing dishes, Schwindel held a soft-spoken sway on his immediate audience on a variety of incidental topics. Several such conversations by groupings were going simultaneously.

"We must avoid the mistake made by Archbishop Romero in El Salvador," Schwindel stated. "Romero was shot because he was wrong. He said that Christianity has a mission to end cruelties and repressions of peoples. What he failed to understand is that it is of the essence of charity, not Christianity, to address the sufferings of people." Schwindel's theme was soon joined by others.

Cardinal Wiggins, prefect of the Congregation for Divine Worship, a graceful man with handsome face, jumped in. "The idea that it is the Church's role to free men from poverty is nonsense. The Lord told us we would always have the poor with us. It is clear that God has given the poor to us that we might practice the virtue of charity."

Cardinal Luigi Peroni's egg-yolk eyes moved around the table. "If you look below the surface you will find that all these demands for social change are fostered by the liberals and the communists. All they do is promote hate. That alone is enough to prove their error."

Schwindel took center stage again. "When you give the poor a false sense of value, they are soon demanding more and more." The prelate's supple fingers played the table surface with audible taps. "One cannot blame people of consequence for fighting to maintain their property and status." The cardinal sliced a piece of his veal bolognese and delivered it to his mouth.

The conversation confirmed Frederick Patrick's esteem for important people. But, at this table it was the men themselves who promoted his inner unease. Where is the humor? Where is the *joie de vivre* he assumed went with prominence?

Toward the end of the meal, Frederick Patrick, silent heretofore except for nods, noes and yeses, ventured a comment to Schwindel. "You should be applauded, Your Eminence, for your skilled handling of the Tubingen matter." A Canadian prelate, questionable by Vatican standards for consistent alignment with Vatican judgment, had been offered an honorary doctorate by the University of Tubingen. The award had been successfully blocked, credit for which had been publicly awarded Schwindel.

Schwindel stared at Frederick Patrick. "It is not a happy memory, Bishop."

Della Tevere heard the squelch and understood. Dispassionate rebuke was another exercise by which Schwindel needled his veins with his favorite drug.

138

The dinner over, the group moved to an adjoining study for coffee and cordials. Frederick Patrick mingled at the corner of the room, out of the grand inquisitor's reach.

*   *   *   *

"Federico, I heard your conversation." Della Tevere's early morning call had wakened Frederick Patrick. "Do not worry about Ludwig. If you had not given him the opportunity to speak down to you, he would have invented one. As it is, you did him a favor and he is in your debt. He will not forget, and will be looking for a way to make it up to you." Before the bishop could respond, della Tevere continued, "You will leave for Firenze day after tomorrow, is it not so?"

"Yes, Alberto. And thank you again for your gracious hospitality."

"It is nothing. We are friends. Enjoy Firenze. We will be in touch, yes?"

The sleek, silver Rapido pulled out of the station, wound its way out of Rome, and sped through the Italian countryside toward Florence. Frederick Patrick sat comfortably in his first-class coach, saying a prayer of thanks for a chance to relax.

CHAPTER EIGHTEEN

"So, how's the tomatoes this year Father Frosty?" A smile broke onto the whey-colored face of the man in bib overalls.

"Why, they're wonderful, Joseph. I can't believe what a gardener I am. I suppose your tomatoes are the size of crab apples again this year, ya?"

"Not this year, Father. This year I got you beat. My tomatoes are like grapefruits." Joseph's eyes were bright. He held up a tomato. "Look at this one."

Frosty reached out and took the tomato in his huge hand. "You call that big, Joseph? Hah! I'll show you what's big. You just bring the wheelbarrow with you tomorrow and I'll give you one of my tomatoes."

So went the chatter every morning after weekday Mass outside the church of St. Francis of Assisi. Gertrude and the other dozen or so attendees all stopped outside the church to visit. Frosty, in cassock, joined them. The white wooden church with its single white wooden steeple beaconed from a hilltop. The slope to the west held the cemetery. Burnside, a stew of sheds and houses for seven score of retired farmers and folks who worked in John's Point, ran along the hill, south and southwest. On the north side the slope carried to a woods. The remaining slope fell eastward, lawn covered, until it formed a shore on Bear Lake. Decades ago the hydroelectric company had purchased the valley. It then built a dam into the Bear River south of the village, and the valley had filled with water. Willows now drooped fingers to the ground on the lake's rim in Burnside. The remaining shoreline was wooded, protected from development by the farmer owners.

Gertrude excused herself from the group to go and start breakfast. The white frame, two-story rectory stood thirty paces to the north of the church. The group dissolved within minutes and Frosty walked to the rectory. The kitchen smelled of fresh-baked bread. Gertrude had made the coffee and was busy preparing the food. The breakfast nook, extending the kitchen, looked out a spacious bay window to the lake and the woods.

"Don't you think, Gertrude, that I could have bacon and eggs more than once a week? I'm seventy-nine years old, ya?, and they haven't killed me yet." Frosty's brown eyes softened when she placed his plate in front of him.

"No, Monsignor, you shouldn't even have that much, so says the doctor." Gertrude untied a flower-decorated apron from her trim waist, put her own plate on the table and sat down. Frosty led the grace and dug into the American fries. "You be satisfied or I'll put you on oatmeal every day." The good-natured threat had been effective down the years. She looked across the table at Frosty. He is in good health, she thought, and we're going to keep it that way. "Maybe on the anniversary, you'll get bacon and eggs, special."

"Ya, Gertrude, just think. Next week you will have been with me forty years." He recalled the time and the slight, pretty young brunette who answered the ad he had placed in the St. Boniface parish bulletin. Fresh out of high school, she was the oldest child of nine in an Austrian farm family. He smiled at the memory. Dark eyes downcast she had professed a lack of interest in boys, clearly with the thought it might help her win a job in a celibate world. In truth, there were not many young men available in wartime.

"It has been a good forty years, Monsignor." It truly has been good, she thought. The memory of that one time came back, as it did on occasion. A decade into their relationship,

Frosty, excited from drinking too much brandy at a priests' gathering, had come home and suggested sex. "I do not think so tonight, Father," she had responded. "But, if you still want me tomorrow, I will." The next day an anguished Frosty had apologized and thanked her profusely for her wisdom. Then and there he had wrapped her securely in the memory of his guilt. There had never been another incident. "You should wear a hat in the garden today," Gertrude said, smiling at the old man. "The sun will be hot."

"You know I hate hats, Gertrude. I hate beets, and I hate oatmeal, and I hate hats. That way I don't have any hate left for people. So you wear the hat for both of us. Besides, this morning I will visit the hospital and our homebound. The sun will not be so high when I get to the garden. I'll bet you will be with your flowers on such a beautiful day."

He looked out the window at Gertrude's domain. Flowers surrounded the church and rectory. Rainbows of flower beds cascaded down the lawn to the lake. Purples, reds, corals and whites encased the shrine of St. Francis between the church and the rectory. Their fragrance filled the rectory when winds were gentle. Gertrude had even repopulated the woods with jack-in-the-pulpit, trillium and lady slippers. And on the moist fringes of the lake, at lawn's edge, purple gentians waited for their autumn glory. Gertrude's flowers were her vigil lights, prayers that kept rising as long as the wick or the bloom flamed.

"I will cut fresh ones for the church and rectory today, and I will take some to Mrs. Schultz. She has not been good since the pneumonia. Will you have lunch at the hospital when you're there? I could send some oatmeal with you." She giggled.

"Ya, I will eat there. Bacon and eggs twice in one day. Don't you worry about me, Gertrude," he teased back.

143

By midafternoon, Frosty was in overalls. He checked the progress in his garden. "You are my roses and my petunias," he talked to the vegetables. He called it his summer garden. "And you," he said, looking at the lake, "you are my winter garden, my soup pot. I do pretty good with the both of you."

He was hoeing into the weeds between the potato rows when he heard the footsteps. He looked up to see the black suit and roman collar. "How are you today, Monsignor Higgins? You have picked a beautiful day to visit us here in paradise," Frosty greeted the chancellor. "What brings us this surprise?"

"Good afternoon, Monsignor. I had business in John's Point at Immaculate Conception parish. I tried calling from there but there was no answer. I guessed that you and Gertrude would be in the garden so I took the chance and drove out. The bishop asked me to come and see you. Can we talk inside? It's very hot out here." His chubby face was red and moist.

"Sure, Michael. Come, we will go inside. We do not often get such a distinguished visitor." They walked up the hill to the rectory. Higgins waved at Gertrude, busy with her flowers. Inside, Frosty led Higgins to the parlor. Frosty tried to recall either a call or a visit from a chancery official that had ever boded well. There were none except for the call from Bishop Zimmerman to give him the news of his promotion to monsignor. It meant something in those days. Today, nichts!

Seated in the parlor, Higgins moved at once to his mission. "The bishop asked me to discuss several matters with you, Monsignor." Higgins skirted first names as though they were septic. "The first two have to do with pastoral matters."

"Ya, vell, shoot, Michael."

"First of all, one of your parishioners wrote the bishop and was absolutely scandalized at your recent ecumenical service. Is it true that you actually had Protestant ministers say some of the prescribed prayers at the holy sacrifice of the mass?"

"I did that, Michael." Frosty looked warily at Higgins. "It seemed the right thing to do. Is it so bad to pray together?"

Higgins emitted a disapproving cluck. "Next time, Monsignor, the bishop wants you to check with the chancery first."

"OK, I can do that." Is this all you chancery guys have to do? he thought. Frosty plucked a sandburr from his overalls and then continued his careful attention.

"The second matter is more serious, Monsignor." Higgins lowered his voice as if to quarantine a contagious disease. "It has been reported to us that you continue to hold penance services in which you give general absolution. I'm sure you're aware that the rubrics allow those services only for the forgiveness of venial sin and do not permit the full absolution of the sacrament of penance."

"I guess that's true, Michael." Frosty knew the rules, but thought they were foolish. He couldn't see the sense in getting people in church and telling them: "We are only forgiving venial sins here tonight. All you folks with mortal sins must go to separate confession, which will be held immediately after the service." Did the bishops seriously think anyone would stay for confession after that?

"The bishop directs you to stop that practice, Monsignor. You are not wiser than the Church."

"I don't want to argue the matter, Michael. If that's what the bishop wants..."

Higgins interrupted. "I'm not here to argue either, Monsignor. I've advised you of the bishop's directive."

"OK, Michael. I got the directive. What else you got?" I wonder who they think sacraments are for, he thought. Are the people to take second place to the rules and rubrics?

"The final matter has to do with parish assignment, Monsignor." Higgins voice became conciliatory. "You have been at Burnside for fourteen years now. The bishop thinks it's time for you to take another assignment. He and the personnel committee ask that you accept an appointment to St. Leo parish in Arkdale. They feel the load there is lighter."

"Michael," Frosty's stomach began to churn, "I am grateful that the bishop is so thoughtful about my workload." This should not be happening, he thought. He had settled this years ago with Bishop Zimmerman. He could picture the bishop nodding agreement to the conditions for his coming here. "Perhaps the bishop does not understand. You should understand, and please tell the bishop, that I came here to St. Francis parish when I was sixty-five years old. Don't you remember what a mess this parish was? Alcoholic priests so needy themselves, no time for the people. The people had been neglected. Can you believe that, Michael? Many priests refused to come here."

Frosty paused to fill his lungs. "Bishop Zimmerman asked me to come here. I said OK. But no more moves, I said, because I'm too old. He promised me, if I come here, I could stay until I die." Frosty stopped to catch his breath. He felt his heart racing.

Higgins remained silent.

After a moment Frosty continued. "So you go back, please, and tell Bishop Sweeney I am grateful for his concern, but the workload here is not too heavy for me and I will stay here at St. Francis as I was promised."

"Very well, Monsignor, I will convey your answer to the bishop." Higgins dropped the subject.

"By the way, Michael, who is it that wants my St. Francis parish so badly, if I may ask?"

"I'm not at liberty to answer that, Monsignor." Higgins got to his feet.

"Ya, vell then, Michael, give my regards to the bishop." He saw Higgins to the door. Frosty's belly churned for days at the recall of Higgins's message.

A week to the day of Higgins's visit, Frosty walked back toward the rectory from his roadside mailbox, scanning through the mail with a careless shuffle. When the chancery's return address popped into view, the churning started all over. Once inside, he slit the envelope:

Dear Monsignor Oberkirche:

> *Monsignor Higgins has faithfully reported his conversation with you of August 3. We have diligently searched your personal file, the file of St. Francis parish, and our own chancery archives. We have been unable to find the agreement you claim to have made with Bishop Zimmerman.*

> *Accordingly, I am appointing you as pastor of St. Leo parish in Arkdale, the appointment to be effective on Monday, September 5.*

> *We deeply appreciate your many years of service for the Church at St. Francis parish. I am confident you will continue your dedicated labor at St. Leo parish. In Christ's gracious charity, I remain*

Yours truly,

Frederick Patrick Sweeney

Bishop of Mill Valley

Frosty dropped into his desk chair. Beads of sweat formed across the expanse of his bald head. Do they think I am a liar? He asked himself. Would I invent such an agreement I had with Bishop Zimmerman back in those days? After all these years they would do this to me?

That night Frosty slept fitfully. He dreamed he was standing at the altar steps in St. Francis Church. He watched the elephant, with a miter on its head and staff in its trunk, slowly lumber up the main aisle of the Church. Frosty tripped backwards under the mammoth's pressing advance and watched in disbelief as the animal stepped on his chest. He raised his head and called out, "Gertrude", but his smothered lungs could force no sound. His head fell back.

Gertrude found him when he failed to appear for Mass.

CHAPTER NINETEEN

"He did not have to die, Father Charles. It is so unfair." Tears streamed down her face as Gertrude and Charles looked out from the rectory window at the crowds of people. Three and four abreast they formed lines that snaked the single block length of St. Francis Street up the hill from Burnside. Many held rosaries as they waited their turn to enter the church. The non-reserved pews were long filled and the people now filed up the main aisle to the coffin, paid their silent respect to Father Frosty, filed down side aisles and departed the church. They crowded back onto the street and lawn, and waited.

Mostly they were older people. They came from towns and villages where Frosty had served, St. Gregory's at Fox Marsh, Holy Martyrs at New Munich, St. Boniface at Ilanz, and more. They talked, each with his own story of where Father Frosty had met their lives, and how important that meeting had been.

Frederick Patrick Sweeney waited impatiently in the living room of the rectory, fully vested for the funeral Mass. The Mass had already been delayed nearly an hour. Gertrude had blocked the procession of bishop and priests when they had started toward the church. "These people come a long way," she had pleaded. "If we got no room in church for them they should at least see him one more time." No one challenged her. Resigned, they simply waited until the last mourner had approached the coffin, signed himself with a cross, gazed at the body, and let personal memories float behind teared eyes.

Frederick Patrick was the celebrant. Monsignor John Haggett, pastor of the cathedral in Mill Valley, served as deacon, the chancellor, Higgins, as subdeacon. Members of the clergy sat in reserved seats and filled a third of the church.

Charles Mueller preached the homily. He talked about *once upon a time* when Monsignor August Peter Oberkirche, Father Frosty, had performed the deeds that grew over time to legend. There was the time "Father Frosty spent with the farmer in Fox Marsh. The man's wife was deathly sick and the doctor, who made farm calls in those days, was unable to diagnose the illness. Father Frosty moved in. He helped the man with the chores. He helped care for the babies and youngsters. And he prayed. He didn't leave and he didn't stop praying and taking care of kids and animals for three days. On the third day that farmer's wife made a turn for the better. That lady is here with us today."

There was the time "when Father Frosty walked into the home of a disturbed parishioner in New Munich. The man was armed with a rifle. His wife had fled the house with their children and alerted the sheriff. The wife pleaded with the sheriff not to hurt her husband. When Father Frosty came, deputies had already surrounded the house. The man fired his rifle several times into the air to keep them at bay. Father Frosty walked up the path to the house. Bullets puffed the dry soil around his feet, but he disappeared into the house. An hour later he came out with the man at his side. As it turned out, a miscalculation in the man's medications had caused his temporary madness. When the people called Father Frosty a hero, he dismissed it. 'Ach, it was nothing,' he said. 'I knew the man was a good man.'" Charles paused. He watched as heads nodded. One old man raised his hand with a finger pointing to himself. "It's true," he said aloud. "I am that man."

"My dad tells this story," Charles continued. "I was only two years old at the time. We had a blizzard in Ilanz. The call came in to Father Frosty. A lady was dying. She lived about six miles from town. The roads were closed to cars because of the heavy snow. Father Frosty took a horse and sleigh, borrowed from the blacksmith, and set out for the farm. He got there. He said he used a compass. The lady died the next morning, but not before receiving the sacraments from Father Frosty."

Charles paused and looked at the congregation. "That lady was my grandmother. She lived with us." He paused again. A smile crossed his face. "Now, there are some tales told about Father Frosty that may be a bit exaggerated. For example, my dad says Father Frosty was pulling the sleigh when he got to the farm. And, the dog-tired horse was a passenger. Now, my dad, who is sitting right over there," Charles nodded to the left, "is one of your better storytellers." The congregation tittered.

Charles had the audience in his grip. "Father Frosty was my hero. I am a priest today because of his example. The most important thing in the world to Father Frosty was the person who faced him. Whoever that individual was, he or she had his total attention."

Charles knew that clerical gossip flies swiftly in its closed circuit, faster through a diocese than local gossip through a town the size of Burnside. Each member of the clergy present knew the events that preceded Frosty's death. Each knew the member who coveted the parish, who had so importuned the bishop with pleas and hints of resignation, and who served at the bishop's right side as deacon at this mass. His closing remark would have a more specific meaning for them than for the parishioners.

"We should never forget the legacy of Father Frosty. Each person he met was Christ to him, at that moment, the most important person in the world. We can give him no

151

greater honor than to let him live on in our lives by imitating him. We must never put anything, our jobs, the organizations we work for, even the revered institution of the Church, ahead of the individual person. To love one another, that's what Christ and Frosty are all about."

They buried Frosty in the cemetery down the hill from the church. The people who waited outside the church had already walked there before the procession left the church. They stood, circling the grave site. The excavated dirt lay mounded, covered by a green carpet, and the bleak tube skeleton of the casket's lowering device spoke its silent finality. Some huddled aside to give the pallbearers and ministers a lane to the grave, and then closed in again. Frederick Patrick took the opportunity to lament "the terrible loss the Church has suffered with Monsignor Oberkirche's passing. The Church is suffering for vocations. Who will replace Monsignor Oberkirche if you good people do not send your sons to the seminary?"

Charles helped the bishop unvest in the rectory after the ceremony. As Frederick Patrick removed each vestment Charles took it and hung it up. Unvested, the bishop turned to him. "I thought your remark that individuals take precedence over the Church both naive and simplistic, Father Mueller. What do you know about running a diocese?"

"Perhaps very little, Bishop." Charles looked the bishop in the eye. "But, does naive and simple make it wrong?"

CHAPTER TWENTY

Stanley Wright and John Winslow Alexander waited at one side of the sacristy and watched the bishop and priests pull long white albs over their bodies. Monsignor Higgins hefted a long bejeweled cope onto the bishop's shoulders.

Excited and restless, Alexander turned to Wright. "To think that we are about to be initiated into the vast historical grandeur of the Church! Certainly makes my heart pump a little faster. How about you, Stanley?"

Wright worked a look of boredom from his face. "Absolutely, John, absolutely! This is quite an honor the bishop is giving us here."

Bishop Sweeney walked over and joined them. "The procession is about to start, Gentlemen. Are you ready?"

Trumpets signaled the processions's entry. The trumpets were followed by organ blasts that reverberated through every corner of the old cathedral. A full choir gave voice to the majestic *Te Deum* as the procession toed a slow pace down the center aisle. A train of surpliced clergy followed candle-bearing altar boys and a cross-bearer. Two acolytes came next bearing crimson pillows. On each pillow lay a long slender sword and a gold medallion on a chain. John Winslow Alexander and Stanley Wright followed next in single file. Each wore a jacket of dark green, embroidered front and back with oak leaves. Their trousers were white with silver side stripes. An empty scabbard hung from their waists. Each man carried a bicornered hat. A robed and mitered Frederick Patrick Sweeney, staff in his left hand, blessed the crowded pews as he followed the procession.

The bishop's voice replaced the rich resonance of the pipe organ as he began his homily. "My dear friends in Christ. The Order of St. Gregory the Great was established by Pope Gregory XVI in memory of the first pope to bear that illustrious name. He did so to honor men of his own Papal States for exemplary service. The honor has since been extended throughout the world. It is our privilege today to confer that great honor, on behalf of His Holiness, the Pope, on two of Mill Valley's outstanding residents." He paused for effect. "You will soon greet *Sir* Stanley Wright, Knight Commander of the Civil Divisions of the Order of St. Gregory the Great, and *Sir* John Winslow Alexander, Knight of the Civil Divisions of the Order of St. Gregory the Great. Nobility has come to Mill Valley." The bishop turned to look at the two knights in waiting. Wright and Alexander knelt ramrod straight on separate prie-dieux in the middle of the sanctuary.

"The days of knighthood are not dead," the bishop continued. "The days of chivalry are not over. Mercy and charity continue to walk our streets in the noble steps taken by men such as these two we honor today. These talented men have literally showered our fair community with their largesse, their talent and their charity."

Alexander's head tilted upward. He sniffed.

Wright's shoulders pulled back.

"Such are the character and courage of these two men." Frederick Patrick concluded his list of Wright's and Alexander's attributes and achievements.

Bishop Sweeney approached the two kneeling men. Taking a sword from its pillow he tapped Wright on the shoulder with the sword's tip. "By the power invested in me by the Holy See, I dub thee *Sir* Stanley Wright, Knight

Commander of the royal order of St. Gregory the Great." Turning to Alexander he again tapped a shoulder. "By the power invested in me by the Holy See, I dub thee *Sir* John Winslow Alexander, Knight of the royal Order of St. Gregory the Great. Arise, Sir Knights." Both men stood and the bishop presented them with swords and medallions. They turned to face the congregation.

"My dear friends in Christ," the bishop continued, "let me describe the decoration itself. It is a red enamelled gold cross. In its center is a round blue enamelled medallion. On the medallion is impressed in gold the image of St. Gregory the Great. His name in gold surrounds the medallion. On the opposite side are the words, Pro Deo et Principe, for God and the Prince. It is magnificent. Be sure to examine it at the reception."

Charles Mueller sat, without cassock or surplice, in the rear of the cathedral. He watched the procession leave the church. His head shook to the rhythm of his silent thoughts. Lord, what has this to do with you?

* * * *

"Listen. You can hear the cathedral bells. How come you're not there, Bill?" Eileen asked. A bright sun warmed the September Sunday. The Valley Cafe had its windows open. Maple trees outside were beginning their fall color change.

"Where?" Bill asked.

"At the cathedral for the investiture ceremony. I'll bet Josephine is excited. She so loves all that pomp and

155

ceremony. But, they're your friends too, aren't they?" She took a sip of oolong tea.

"Friends is probably a little strong, Eileen. I know them well enough from all those years at the bank. Since I retired, though, I've been out of the loop, their loop I mean. Don't see them much anymore. Oh, I see Alexander at the college now and then, but that's about it."

"They should be making you a knight, all the good you've done for this community over the years."

"Nonsense." Bill said it gently. "Bishop Zimmerman offered it to me once. I turned it down. Can you think of anything more useless or out of touch with reality?" He changed the subject. "I appreciate your help today, Eileen. I'd take a sandburr on my butt any day over shopping for clothes. Mary always kept me out of trouble. She'd inspect me every day before I went to work. More than once she stopped me from walking out colored like a circus clown. Now, I've had to put my own system together."

"I admire the way you deal with it, Bill."

"Eileen, do you mind if I ask you a personal question?"

"What kind of personal question?"

"The kind bankers ask?"

"I don't follow you."

"How are you fixed financially? Do you have what you need?"

Eileen paused for a moment. "I'm getting by just fine, Bill. I've built on Tim's insurance money and I've saved over the years."

"Did the diocese have a retirement plan for you?"

156

"No."

"I might have known without asking."

"Now, Bill, let's not judge," Eileen calmed.

"Eileen, I've got enough money for a dozen people. It would make me feel wonderful...I would be honored...if you ever need...I just don't want you to need."

Eileen reached across the table and patted his hand. "Thank you, Bill. If I ever need I'll come to you for help."

"Promise?" His eyes moistened.

"Promise."

CHAPTER TWENTY-ONE

The reception sparkled. High profile guests mingled, carrying liquids in crystal and plucking exquisite hors d'oeuvres from silver trays carried about by uniformed waitresses.

"Congratulations, Mr. Wright." Maggie shook Wright's hand.

"Why, thank you, Sister Margaret. And, please call me Stanley. After all, we'll soon be working together."

"I'm looking forward to that, Stanley. It was kind of you to accept a position on the board. Your reputation precedes you so I know what an asset you will be for Padua College. We need the presence of good business minds."

"Well, thank you. I enjoy supporting the Catholic enterprises of our area. In fact, I'm dedicated to it. And certainly, Padua College stands out as one of our most visible Catholic institutions." Stanley noticed other guests waiting for his attention just beyond the reach of their conversation. "I should greet these other guests, Sister Margaret. I must say, though, I look forward to learning much from you about the college. The board meets this coming Tuesday evening, do we not?"

"That's right, Tuesday at seven o'clock."

"Nice of you to help me celebrate my knighthood, Sister. Now, if you will excuse me, I will look forward to seeing you again on Tuesday. Enjoy yourself." He was away.

On the other side of the Country Club's ballroom, Frederick Patrick Sweeney loomed above John Winslow Alexander. "Well, now, Sir John, how does it feel to be a royal member of the papal household?" The bishop raised his cocktail and clicked the glass of John Winslow Alexander.

"Your Excellency, my spirit is soaring. I feel simply intoxicated. St. Peter could say of me what he said of the Apostles on Pentecost. This man 'is not drunk. It is too early in the day.' I thank you again for this great honor."

"You are most welcome, John. All of the people I talked to were unanimous in their endorsement of you. They only confirmed my own judgment. Your influence has spread beyond the walls of Padua College. You are as well-known for your catholic values as you are for your academic stature and philanthropy. And it has given me great pleasure in promoting this recognition for you. Now tell me, how are things at Padua?"

"Gentlemen, your glasses are nearly empty," Stanley Wright noted as he joined the twosome. "Waitress, take orders here, please." Wright, footing the bill for the reception, played the solicitous host. "May I join this conversation?"

* * * *

On Tuesday evening, Phillip Wolfe called the meeting of the Padua College Board of Trustees to order. Wolfe, an industrialist, obviously enjoyed the presidential prestige given him by his peers. His welcome to the new members, Stanley Wright and Samuel Stein, ended with polite applause for the newcomers. The formality of the annual elections followed a prearranged agreement of the Executive Committee members. Wolfe was quickly re-elected president. He then ran through his list of committee appointments, a confirmation of the status quo. New member

Wright was appointed to the Executive Committee, responsible for day-to- day oversight of the college and for the handling of *unusual events*. Stein went to the Finance Committee.

"Sister Margaret has done her usual competent job on our agenda," Wolfe announced. "She will brief us on our agenda items. The first item is the construction of the new theatre building."

"Thank you, Phillip," Maggie responded. Let me briefly review the history of the theatre project for the benefit of the new board members. The board first approved the project three years ago. Since then we have raised the funds necessary for construction. In fact, we have exceeded the goal of four million by 150,000, thanks mainly to the generosity and efforts of the board members. Two years ago, the board selected the architectural firm of Schlum, Filbert and Sartori. Plans were drawn, discussed, and revised. They were approved by the board this past spring.

"We have since put the specifications out for bids, having the architects provide oversight to the bidding process. They have reviewed the bids and the competency of the bidders. They have endorsed the competency of both the Wright Construction Company and the Thompson Construction Company. I am now able to give you their recommenda-tion..." Maggie paused and looked at Wolfe.

"Excuse me, Mr. President," Stanley Wright interrupted. "I think it appropriate that I absent myself not only from the vote on this issue, but also from any further discussion. I only want to assure the members that my enthusiasm for continued service on this board will not be affected by your decision in this matter."

"As you wish, Stanley. I was about to suggest the same course of action," Wolfe replied. "We will find you in the faculty lounge when we have disposed of this matter." Wolfe was matter-of-fact. Wright left the room.

"Given the competency of both the Wright and the Thompson companies,' Maggie continued, "the architects recommend the contract be let to Thompson based solely on their bid. The Thompson bid is $90,000 lower than the Wright bid." Maggie turned the meeting back to Wolfe.

Maggie watched as Wolfe adroitly orchestrated a dance of discomfort among the trustees. Would not a contract with Thompson imply a lack of appreciation for a staunch Catholic?

What is $90,000 anyway? When the discussion ended, a contract with Wright Construction was approved on a vote of ten to seven. On his return to the meeting Stanley was profuse in his gratitude for the vote of confidence.

At a signal from Wolfe, Maggie continued her briefing: "The staff of our Employe Assistance Program have made an intervention, with my specific approval, on one of our faculty members for alcoholism. The performance of Professor Simon Schaff of the Chemistry Department had deteriorated to an unacceptable level. On investigation, the staff discovered the alcoholism. I gave Professor Schaff the options of rehabilitation or termination. He accepted the rehabilitation and is now on administrative leave."

"Is the administrative leave with or without pay?" Wright asked.

"It is with pay until his accrued sick leave has been exhausted. After that, it is without pay," Maggie replied.

"You should have fired him on the spot, Sister Margaret," Wright asserted.

"Professor Schaff is an excellent teacher, Stanley, and a good ethical man. He is sick. I think the college can support his rehabilitation efforts. It's the Christian thing to do."

"Believe me, Sister Margaret, you're just asking for more trouble down the road. I don't see this as a matter of being Christian or not. It's business. I wonder about your judgment on that," Stanley concluded.

"Sister Margaret," Phillip Wolfe cut in. "What has been the impact of the May streaking incident?"

"None that I can detect, Phillip. I believe it was an isolated event. Not many students were even aware of it."

"What will parents think of Padua?" asked Wolfe.

"Doesn't the incident portray a lax attitude on the part of the administration?" asked Wright.

"Are you losing control of the students?" from Wolfe.

"Have the streakers been punished? What was the punishment?" this from Wright again.

Maggie raised her arms, palms out. "Halt already." She smiled. "I can only handle so many questions. Parents were once students themselves. We have had no complaints from them. As for the administration, our primary role is to provide the opportunity for a good education. The provision of an arena of freedom to students is required in that process. With freedom, incidents are bound to happen. No amount of control will prevent incidents. We deceive ourselves if we think a college environment can be entirely controlled. So, we are not in the business of control. Finally, there has been no punishement as the streakers have not been identified."

"I don't mind telling you, Sister Margaret," Wright said, "I'm bothered by that attitude."

CHAPTER TWENTY-TWO

Tony Corsalini walked into the yard at the Wright Construction Company. Cranes and shovels loomed over dozers and crawlers, silhouetting an eerie Jurassic dawn. Metal sheds joined by an asphalt network dotted the yard. Gale force winds shocked heavy metal sounds against the sheds. Hard stinging snow pelted his face as he moved from shed to shed where he checked assignments and schedules, men and materials, with the foremen.

The sheds were shops, independent business entities under the Wright Construction flag: the roofers' shop (Wright Roofing, Inc.), the carpenters' shop (Wright PreFab, Inc.), the HVAC shop (Wright Heating, Ventilating and Air Conditioning, Inc.), the plumbing shop (Wright Plumbing, Inc.), the electrical shop (Wright Electric, Inc.), the diggers' shop (Wright Excavating, Inc.), and the masons' shop (Wright Masonry, Inc.). Each of them took direction from the office. Each of them worked on two levels. On one level they operated jointly on the big projects of Wright Construction. On the other, they worked on independent projects of their own. Scheduling so that the two levels balanced efficiently was the work of Tony Corsalini.

Long weekends are fine, Tony thought, but it's a pain picking up all the strings on Monday morning. Back in the office he sloughed off his jacket and wool stocking cap and hung them on the rack in the cubicle he shared with Stuart Wright. He walked back into the reception area where the coffee he had started before his rounds was now fully perked. He had both hands absorbing warmth from the cup when Stu walked out of Stanley Wright's office. "Hi Stu, how was your Thanksgiving?"

"Ate too much again," the younger man replied. "Do it every year. You'd think I'd learn." The thin Stu looked as if the stomach ache was still on him. "How about you, Tony?"

"Real nice. Kids, my parents, Kate's parents. Kate did the turkey thing. Watched the Jets beat the Packers on Sunday. Thank God that strike is over. I missed my football."

"Tony, can I see you for a minute?" Stanley Wright stood in his office doorway.

"Sure, Stanley." Tony followed Wright into his office.

"Close the door, will you, Tony?" Tony closed the door and took a chair in front of Wright's desk.

Wright seated himself behind his paper strewn desk and leaned back in his chair. Behind him a large oak roll-top served as credenza and back drop. "Tony, I've decided to let you go." Wright's face was unperturbed, his voice quiet.

"I'm sorry, Stan?" Tony's face registered confusion.

"I'm letting you go. I'm terminating your employment with Wright Construction." Wright's eyes were impassive.

"You mean I'm...I'm fired?" Tony leaned forward, eyes wide, incredulous.

"That's correct, Tony. It's effective immediately. I want you to clean out your desk and leave. Stu will take over your job."

"But why? My God, Stanley, I've worked hard for you. I've more than earned my keep. You said so yourself."

"The decision's made. I don't discuss my reasons, Tony. Wright stoically tapped a pencil on the edge of his desk.

"I have a family to care for. God Almighty!"

Wright was silent, looking directly at Tony.

Tony became silent. His wild eyes betrayed his spinning thoughts and pounding heart. "But why?" he asked again.

Wright remained still.

"It's not fair, Stanley," Tony said, almost pleading.

Wright said nothing.

After some time, "I've heard there's a job at Thompson Construction. What about references? Can I count on a good reference from you, Stanley?"

"I'll do what I can, Tony. Frankly, I think it would be in your best interest if you looked for a job away from this area. An ex-priest is bound to get some local opposition."

"What does that mean? You knew that when you hired me."

"That's all I'm going to say, Tony. Now, if you'll excuse me." Wright stood up. The meeting was over.

Stunned, Tony left the office and walked to his desk. It took only minutes for him to throw his few desk possessions into a paper bag.

"It wasn't my idea, Tony," Stu commiserated.

The tear-streaked face of Susan, his secretary and the company receptionist, provided the only warmth to his leaving. He gave her a quick hug, threw the paper bag into his pickup and drove away.

* * * *

167

At the rectory Charles lifted the phone to meet Kate's frantic voice.

"Charles, is Tony with you?"

"No he's not, Kate. I haven't seen him today."

"Something's terribly wrong, Charles. I just called Tony's office and Susan said he doesn't work there anymore. She said that he left there before eight this morning. She seemed afraid to tell me any more than that. It's now two in the afternoon. That's over six hours. I'm so worried."

"You stay calm now, Kate," Charles said calmly. "Tony's got the most level head of anyone I know. He's fine. I know it. You said he doesn't work there anymore?"

"That's what Susan said."

"Kate, you stay put with the children. I'll check around and get back to you soon. And try not to worry. I'll find him. I know that scoundrel like a brother."

Charles looked up the number and called Wright Construction, asked for and got connected to Stanley Wright.

"Hello, Father Mueller. Staying warm I hope?"

"I'm fine, Mr. Wright. I have a favor to ask."

"Stanley, call me Stanley. How can I be of help?"

"Kate Corsalini just called me, Stanley. She's worried about Tony. He hasn't come home. Something about not working for you anymore. Are you free to tell me what happened?"

"There's not much to tell, Father. I let Tony go early this morning."

"Let go? You mean, fired him?"

"That's correct."

"That's quite a shock, Stanley. Tony told me only recently how well his job was going and how much he enjoyed it. What happened?"

"I'm not free to go into detail on that, Father. I'm sure you understand."

Charles felt the stone wall in Wright's voice. "You've been helpful, Stanley. Thank you." He hung up and grabbed his coat, cap and gloves.

Three inches of fresh wet snow covered County Trunk P as Charles wound the Chevy along its curves, climbing up the bluff. The car started to slide on a turn. He pulled it out with an opposite turn of the wheel. His speed was down to twenty. Just as he neared the top, the car slid again. This time he could not control the slide. The car mushed into a snowbank and stopped with one wheel hanging free over a ravine. He climbed out of the car and started walking.

At the bluff's summit he turned into an old logging trail. He tried a heel-and-toe walk down the snow covered path of a tire track, but his feet kept brushing the higher surrounding snow. His feet were quickly soaked and he gave it up. He slogged along, one foot in the tire track. At the end of the wooded trail was a small clearing. It looked down on a scene of quiet wonder, an island dotted river that spread its sloughs and marshes for miles in every direction. For years this spot had been a retreat for both men when they needed solitude. Charles saw the truck as he entered the clearing. And he saw Tony's unmoving head through the rear window. He walked to the passenger side of the truck, opened the door and climbed in.

Tony Corsalini sat, eyes vacant. Charles put a hand on Tony's shoulder and relaxed at the body's warmth. "You're going to be OK, Tony."

Tony didn't respond.

"Can you hear me, Tony?" Charles asked gently.

"Yes, I hear you." Tony's voice was lifeless.

"I heard the news."

Tony's eyes filled.

"Kate called. She's worried about you."

"How did she know? How am I going to take care of her and the kids? What about the kids?"

"We have to trust that it will work out, Tony. Things are dark now. Don't try to look ahead while it's dark." Charles's hand rested on Tony's shoulder.

"It's not fair, Carlos. It's not fair." Tony put his face in his hands and leaned into the steering wheel. Tears leaked through his fingers and fell on the steering column.

Charles moved to put a hand on each of Tony's shoulders. He quietly massaged the shoulders until the tears subsided.

Tony lifted slightly and began to pound the dash with his fists. "Damn! Damn! Damn!"

"Come on, now, Tony. You've got to get home for Kate's sake. Besides, my feet are soaked and freezing. If I don't get to some dry socks you'll be taking me to the hospital. Come on. Let's go." Charles was insistent.

Corsalini looked up. "How'd you get here? Where's your car?"

"In the ditch, a little way down the hill. That old buggy isn't going anywhere. I'll have Spike's Garage send someone to get it."

On the way to the farm Tony replayed his firing word by scarce word. Charles watched the reunion with Kate. He felt their tears, and stayed until they were somewhat calm, at least for the moment. It was late evening when Tony dropped him at the rectory.

Once inside, Charles called Maggie.

"This is Sister Margaret."

"Hi, Maggie, this is Charles. I've got some bad news. Stanley Wright fired Tony today."

"Oh no!"

Charles waited while Maggie absorbed the news.

"I feel so bad," Maggie said. "Everything has been so good for Tony and Kate."

"Kate will need some support."

"I know. I'll call her right away. Thanks for calling me."

A burr had been riding Charles's consciousness. As he turned in for the night the burr transformed itself into thought. Who lay behind Wright's ex-priest comment to Tony?

CHAPTER TWENTY-THREE

The candles fashioned a glowing 59 on the cake. Josephine huffed them out. "What a beautiful cake, Eileen. You didn't make it?" Josephine, Eileen and Bill sat around Eileen's dining table.

"No, it's from the bakery, Josephine," Eileen replied. "But, Bill and I did make you something."

Bill walked to a closet and returned with a large rectangular package, carefully wrapped in bright pink paper, and topped with a large blue ribbon and bow. He began to sing off key, "Happy birthday to you..."

Eileen joined in with her husky voice.

The song finished, Josephine removed the ribbon and bow and carefully set them aside. "Such a pretty pink," she exclaimed as she removed the wrapping. She stared at the likeness of herself. "It's beautiful, Eileen. It's so flattering. I...I look so good!"

"Nonsense, Josephine, it doesn't do you justice."

"It's good, all right, Josephine, but Eileen's right," Bill agreed.

"Didn't Bill do a good job on the frame?" Eileen asked.

"It's...it's just wonderful," Josephine stammered. I don't know where to hang it. Maybe in my bedroom?"

"It goes in your living room, Josephine," Bill directed. "On the wall directly in front of the entrance."

"Yes, it will fit there, won't it?" Josephine was grateful for the solution. "Thank you both."

"Come on now, it's time for cake," Eileen said. "Did I tell you that the Holy Spirit is sitting for my next portrait?"

Brunch over, they played cards.

"Did you hear that Stanley Wright fired Tony Corsalini?" Josephine presented the news like a jeweler offering a diamond on a velvet tray.

"Oh no! How awful! How sad." Eileen's face read her heart.

"I don't think it's sad at all," Josephine responded, her thin nose up.

"I remember how pleased Bishop Zimmerman was when Tony got that job," Eileen said, glued to her own thoughts. "I feel so bad for Tony and Kate, such good people."

"Like I said," Josephine replied, "I don't think it's sad at all. I never liked him when he was Father Corsalini at the chancery, him or that Father Mueller."

Bill remained quiet. He knew that Sister Margaret was close to Kate Corsalini. He shuffled the cards.

"How can you say that, Josephine?" Eileen asked, surprised. "I've known Tony Corsalini since he was in knee pants. He was a fine boy and he's a fine man. The same goes for Father Mueller."

"He may have been a nice boy, Eileen," Josephine's golden coif shook with her reply, "but he abandoned his priesthood. And that says it all for me."

"Are we going to play pinochle or what?" Bill was calm but insistent. "Your bid, Josephine."

"Just a minute, Bill." Eileen reached out to cover Bill's fingers with her hand. "Josephine, I'm surprised to hear you speak like that. How can you make such a judgment?"

"Easy, like I've told you so many times, Eileen. God gave me this *gift*." Josephine's gift had passed the same judgment on both Tony Corsalini and Charles Mueller. She had first met them at a St. Stephen's parish festival when they were both newly ordained. She had been standing at her place behind the relish dishes, spooning out corn and cucumber and cranberry....She saw who took or didn't take any of Josephine McGregor's famous corn relish. The two priests had passed through the line making complimentary small talk with the women. They failed to take any of the corn relish. To Josephine it was "as plain as the nose on your face," those two guys were nothing but smooth talkers. When Mueller became pastor of St. Stephen's after leaving the chancery, she changed membership to St. Martha's, where the holy Monsignor Deutschman stood firm with tradition.

"Are you so sure your gift is always right, Josephine?"

"I never doubt my gift. Sometimes there's just no reasoning with you, Eileen. We see things so differently." Josephine studied her cards. "I bid fifteen."

"Pass. I haven't had a decent hand yet. Where'd you buy these feminist cards?" Bill tugged things back to level.

* * * *

"I don't know if you've heard yet, Bishop, but I was forced to let Tony Corsalini go last week." Stanley Wright's tone was serious, but wrapped in a smile.

175

The bishop clicked his glass to Stanley's.

It was December 8th, the feast of the Immaculate Conception. Wright was having a party. His home was decorated brightly on the outside with a thousand Christmas lights. The home stood above the city on the side of the bluff, its lights a festive sight enjoyed by those below. Stereophonic violin music gave a soft background to the mix of voices as Stanley's guests mingled. Stanley and his wife, Sarah, had moved easily from group to group until they had personally greeted each of the guests. They then parted and wound separate paths back through the groups, attending to members of their own sex. Their two daughters, Serena and Sandra, both pretty, both college students at Padua, made cameo appearances to greet the guests.

"Your Excellency, Sarah and I are so pleased that you were able to join us. We know how wearing feast days must be for you."

"The quality of your liquor and cuisine is always an attraction, Stanley, second only to the quality of your company." Frederick Patrick lifted his cocktail in salute.

Wright smiled his appreciation. "Also, Your Excellency, I'm sure you've noticed, our guests this evening all have one spouse who is a member of the Padua Board of Trustees. As you mingle I hope you will feel free to breathe a few lamentations on the sad state of the college. Your words would carry a subtle blessing for my next sacred trust."

"I would be pleased to do that, Stanley. If you will excuse me I will be about that task. I have George Murphy and Phil Wolfe in my sights." He made for the adjoining group and began his rounds.

An hour later the lights dimmed momentarily. A uniformed waiter appeared at the doorway and tinkled a small bell. "Dinner is served. This way, please, Ladies and Gentlemen."

The table glistened with crystal and polished silver. A fragrant pine wreath, its red bulbs glistening under the lights, lay at the table's center. Table favors of holly, red ribbon and silver bells added luster to the place settings.

Waiters helped the women to their seats. Stanley led the bishop to a seat at the head of the table, and then stood next to Sarah at the opposite end.

"Your Excellency?" Stanley's head inclined. "Would you honor us with grace?"

CHAPTER TWENTY-FOUR

"Come in." John Winslow Alexander boomed at the closed door of his office. The door opened to the face and muscular figure of Jonathan Meissner. Alexander had told the tale over the years of how he had loved, and lost, the beautiful Melanie Wright nearly thirty years before. "I never found her equal," he would say. The truth stirred in his groin at the sight of Jonathan's blond good looks and supple movement.

"Good morning, Professor Alexander."

"Jonathan, how good to see you. Come in. You're here for the copy?" Alexander stood up."

"Right, Professor. The printer says he wants it by one o'clock this afternoon."

"It's ready to go." Alexander retrieved a manila envelope from a nearby chair. "I reviewed it all last night. My signature is on the cover sheet authorizing its publication. You are to be complimented, Jonathan, on your article, *Romance Revisited.* I must say, you show the parallel between the Romantic Era and our own times with wit, beauty and delicacy. The *Paduan* needs more literary stuff, good prose, good poetry, and less rah, rah, rah for the team, don't you agree?"

Jonathan smiled beautifully but could not hide his sense of haste. "Absolutely, Professor. Thanks a lot. I'll get this copy over to the printer right away. Nice talking with you." Jonathan was out the door.

Disappointment churned Alexander's face into a peevish frown.

The Volkswagen Beetle, hallucinogenic in bright paints, rolled to a stop in front of Birchwood Printing, Inc. Jonathan pried up the metal fasteners on the manila envelope, opened it and withdrew the eight-page typescript. He then replaced pages one and three.

* * * *

Two days later Maggie faced Jonathan Meissner and John Winslow Alexander in her office. "Jonathan, Professor Alexander has informed me that the article in today's *Paduan, 'In Defense of Homosexuality,'* replaced a piece he approved on Romantic Era parallels with modern times. Please explain how that exchange happened."

"I did it, Sister Margaret." Jonathan's eyes were on Maggie's. "I wrote both articles to have identical length, got the one approved by Professor Alexander, then made the switch on the way to the printer." The words were straight out, non-apologetic.

"Why did you do that?"

"It's something I had to do. It's truth to many of my friends. I believe them and I spoke it for them. I knew I'd never get approval."

Maggie was quiet for a moment, looking directly at Jonathan. "Well, Jonathan, you are certainly entitled to your truth. And you are entitled to speak that truth in whatever media either belongs to you or invites it. The *Paduan* is not yours, Jonathan. It is the voice of Padua College. You are entitled to speak there only what is approved by the faculty moderator. Your truth is not the offense here, Jonathan. Your failure to get approval is. I don't have to give a speech here, Jonathan. You know all that."

"Yes I do, Sister Margaret."

"That being the case, you will have to find new media for your truth. As of now you are no longer the editor of the *Paduan*, nor a member of its staff." Maggie turned to Alexander. "I hope that won't place too great a burden on you, Professor."

"We'll manage, somehow, Dr. McDonough." Alexander's voice was subdued.

"I think that covers it. Jonathan, anything else?"

"No, Sister."

Maggie looked from one man to the other. "That will be all then. Thank you both for coming." Jonathan opened the door for Alexander.

The incident gnawed at Maggie the entire morning. She got up from her desk several times and walked her office. She thought to follow up on the already distributed *Paduan* with an explanatory note to the students, but decided that would only give unwarranted notoriety to the deed. She thought to have Alexander do a brief note of explanation in the next issue, but vetoed the idea for the same reason. In the end she decided to do nothing. Deep down she felt a tinge of admiration for Jonathan.

Charles Mueller appeared in her doorway. "That frown doesn't go with the Maggie I know," he said.

"Come in, Charles. That frown goes with a disciplinary action I've just taken. Have you see the latest edition of the *Paduan*?"

"It was in my faculty mailbox this morning," Charles replied. "Haven't read it yet. Why?"

"There's a piece about homosexuality on page three. It wasn't authorized. I disciplined the student this morning, took him off the *Paduan* staff."

"Sounds reasonable. Punishment related to the crime and all that..."

"I have the feeling it's going to turn around and bite me." Maggie shook her head. "Let's change the subject. I got a letter from Kate. They're all settled in. She invited me down for some hot Mexican cooking."

"I know. I got a note from Tony yesterday. He wants me to come down and try a different kind of fishing. He has good vibes about his new job."

"Kate says you made the job connection."

"We all worked hard at networking connections. Tony said that Kate is working part-time at a military hospital."

"Right. Kate says that San Antonio is loaded with them. Is there another reason why you're here, Charles?" Maggie asked, realizing the meeting was unexpected.

"Yes, I've decided to drive down and see Tony and Kate. With Lent and Easter coming up, now's the best time for me to get away for ten days. Jake said he would take my classes next week. So bright and early Monday morning the old chevy and I hit the road. Why don't you come along? Free ride and all that."

"Oh, it would be nice to get out of this snow and cold for some warm sun, but I really can't." Maggie hesitated, then reached for her calendar and studied it. "Yes, I can." She was talking to herself. "A few schedule changes. I could catch a plane from San Antonio for my Seattle talk on the 14th." She looked at Charles. "Yes, I"ll hitch a ride one way with you. You're a tonic, Charles. My spirits are up already."

* * * *

They each wore jeans and sweaters. The snow followed them until they were midway through Illinois. Then islands began to appear in the white. They played games that Maggie invented as they drove: alternately, name a river for every letter of the alphabet, one point for every correct answer. Then, the same for U.S. cities, cities of the world, etc. They played games like *Horses,* two points for a white horse on your side of the road, one point for a horse of any other color. By the time they got to Springfield, Missouri, Maggie led by ninety-four points. Charles conceded.

"Your Seattle talk? Is it the same as the one you gave in Denver?"

"It's about the same, Charles. I'm simply trying to increase women's awareness of the roles they've been assigned down through history in the Church and in society. My hope is that it will help them make free and conscious decisions about their own lives."

"Of course, you ran the talk by Bishop Sweeney."

"Right. Absolutely. I wouldn't breathe a word without prior clerical approval. You know that." She was laughing.

"I like your laugh."

"You're easy to be with, Charles. I think it would be as easy to cry with you as it is to laugh."

"You don't have permission to cry on this trip."

They were quiet for a time. Then Maggie asked, "Do you ever feel alone, Charles? I'm surrounded by community, people I like and enjoy, people who truly support me, yet I'm alone."

"Yes, I have a similar experience. I think it's because we've strayed, the two of us, from the usual thinking of our peers. We started with them, and then something, I hope it's grace, caused us to veer. Now we're alone. We can't go back without doing violence to the conscience that first put us on the road."

"I don't feel alone when I'm with you."

"And I don't feel alone when I'm with you, Maggie. Perhaps God gave us this friendship to help each other travel lonesome roads."

Maggie reached over and put her hand on Charles's arm. "Thanks for that."

They almost made it to Tulsa that day. The night was coal black when they stopped for sleep. "We've come over six hundred miles," Charles said, looking at the dashboard. "With an early start we'll make it to San Antonio tomorrow." The motel clerk assigned them adjoining rooms. Charles carried their overnight luggage. After taking time to wash off the travel soil they met in the lounge. They had a cocktail and then dined in the motel restaurant. The meal was a pleasant surprise and they ate leisurely, letting the wine soften weary travel muscles.

The light from under Maggie's door penetrated the room and woke Charles. He got up, put on a robe and knocked on Maggie's door.

184

Later he could not remember what made him open his arms. He remembered her standing there, her short blue button-down gown, the ripple movement of her breasts, his astonishment that she filled his open arms. He wrapped her in a hug. The hug became a kiss, gentle but firm. The kiss was repeated, gathering length and intensity. Their hands were clumsy. Maggie's hand found Charles's chest as she parted his robe. His hands fumbled with buttons and found her breasts. They fell on the bed.

Suddenly they broke apart, both breathless, floundering.

"We can't!"

CHAPTER TWENTY-FIVE

As they left the motel in the morning, Maggie could see the concern on Charles's face. She watched him as he loaded their luggage and as he drove them onto the interstate. "Are you worried, Charles?"

"Yes."

"About me?"

"Yes."

"I knew it. You think you've ruined me. You think I'm going to jump off a bridge or something?"

"Something like that."

Maggie took in his concerned face. "Have I ruined you?"

"Don't be silly."

"Then let's not either of us be silly. We stopped, didn't we? What we did looks like a big deal because I've never had sex."

"Neither have I," Charles said.

"If I hadn't become afraid, I might have," Maggie admitted.

"Me too. Maybe that's what they mean that fear is the beginning of wisdom."

"And, if I've learned any wisdom from you, Charles Mueller, it's that God looks for a caring heart. I can't believe that an intact hymen interests him one bit. For myself, I can't remember feeling more cared for. I have never felt so much a woman." She paused and smiled at him.

He smiled back. "Would you like to pray together?"

"Yes, I'd like that."

By the time they got to San Antonio late that night they had reached a new sense of intimacy, and their humor had returned.

"So, no more bedroom deviltry, right?" Maggie jabbed his shoulder.

"Right. Besides, you snore. It came right under the door."

"I do not snore, Charles Mueller." This time the jab went to his ribs.

*   *   *   *

Kate answered the phone and turned to Maggie. "It's for you; it's Sister Felicity."

Maggie took the phone to hear Felicity recite the number of calls received from trustees on the homosexuality article. Did Maggie want to call them? No, why incur expense for so many calls? Felicity should call all of the concerned trustees and reassure them that discipline had already been taken. She, herself, would be back at the college in a week and would give a full report to the Trustee Executive Committee at its meeting on February 17.

"Trouble?" Kate asked as Maggie slipped the phone onto its cradle.

"I hope not, Kate. Besides, I'm here to get away from all that. I love your place. It has everything your old farm had, plus a warmer winter. You were lucky to find it."

"Yes we were. Down here it's a ranch. I'm still not used to that word, but I like it. Oh, here's my neighbor to take the girls." Kate looked out the kitchen window. "Come on, I'll introduce you, and then we can saddle up."

* * * *

Tony and Charles carried their fishing rods and tackle boxes from the charter boat, together with twenty or so other disappointed fishermen.

"That's a lot of water out there, Tony. You sure it's got any red snapper in it?"

"Hey, how many years was it before we got out first muskie? We'll get those snapper yet. You can use a new fish, get you out of the muskie rut. Besides, snapper tastes good. You'll see tonight how good they are. Kate's going to fix some."

They climbed into the pickup and pulled out of Port Aransas. It was catch-up time. Tony's new job was a twin to the one he had with Wright. Only here there was the promise of an eventual buy-in.

* * * *

The February 17th meeting of the Executive Committee had the members settled around one end of the board room table. Phillip Wolfe, board president, also chaired this committee. Stanley Wright, George Murphy, James

Sutherland, all businessmen, and Sister Mary Jane Plummer, president of the religious order, were its members.

Wolfe called the meeting down to business by asking Sister Margaret for her report. Committee member comments were sparse as she went through the substantive matters on the agenda. They erupted when she concluded the factual detail of the Jonathan Meissner incident.

Wright adroitly seized control of the meeting. "It appears to me, Sister Margaret, that we are dealing here with another example, albeit much more serious than the streaking incident, of the consequences of the permissiveness which your tenure as president has brought to Padua." From the relative calm of the opening statement, he launched a diatribe against "faggot filth," which he concluded with, "you should have expelled that miserable fag."

Wright's intensity terrified Sister Mary Jane. She placed a hand over her mouth as if the gesture could somehow protect against Wright's temper.

Maggie, surprised at Wright's homophobic reaction, responded first with silence. Then, "Allow me to insert my perspective, Stanley. First of all, Jonathan says the article does not speak his personal experience, but the truth of his friends' experience. He brought the article to me last year asking permission to print it after it was rejected by Professor Alexander. I refused, not as a judgment on homosexuals, but rather as a judgment on what is the purpose of the *Paduan*. Second, the school has set up safeguards to prevent inclusion of inappropriate material. Requiring the signature of Professor Alexander is a reasonable safeguard. I would add that Professor Alexander's signature was appropriate for the material shown him. Third, the material was inserted deceptively by the student. There is no question of that. He has admitted to the deed. Jonathan has been disciplined. He is no longer editor of nor on the staff of the *Paduan*. The insinuation that an atmosphere of permissiveness is responsible for the deception is to deny the freedom and liability of the perpetrator. And, this college is not in the business of exercising controls that violate adult freedom."

"He should be expelled!" Wright was loud and insistent.

"I disagree." Maggie was calm, but firm. "I believe that we must look at the motive here, Stanley. Jonathan's purpose was not an assault on authority. If it had been only that I would have expelled him. He wanted only to make an appeal for understanding. It is that motive that led me to mitigate the punishment. It seems to me that we Christians, and surely a Catholic institution like Padua, should be empathetic to those who appeal for a hearing, who appeal for understanding, especially if we don't share their particular pain. It seems to me that we are called to redeem, not condemn. Jonathan is a very talented student. I have hope that those talents will be productive for society."

Wright shifted uneasily in his chair but said nothing.

Maggie continued. "Discipline is the president's job. This was my call and I made it. And those are my reasons. I trust you will all understand that."

CHAPTER TWENTY-SIX

It was the Wednesday of Holy Week. A cold March rain hurried their steps as they moved individually from the cozy warmth of their cars to the uneven heat of the Padua College administration building. They assembled in the board room, exchanging trite courtesies. Phillip Wolfe called the meeting to order. "Sister Margaret has done her usual competent job on the agenda for this evening. We'll begin with her report."

Maggie said a silent prayer of thanks that the mood of the board appeared light and receptive in contrast to the last Executive Committee meeting. She went through her recital of administrative data, then asked for and received approval on several minor policy changes. Following Maggie's business, committee chairpersons gave their reports.

"We're moving along smoothly this evening." Wolfe looked at his watch. "We may finish early at this rate. Our next agenda item is the annual renewal of the president's contract. Is there a motion from the floor?"

"I move that Sister Margaret's contract be renewed for the coming term." The motion came from Sister Mary Jane Plummer.

"Second," from Dr. Samuel Wasser.

"Mr. Chairman?" Stanley Wright asked for consideration.

"Yes, Stanley?"

"May I suggest, Mr. Chairman, that a matter of such importance should be discussed. I also think it would be appropriate for the trustees to hold the discussion in private."

"I agree, Stanley. Sister Margaret, would you be so kind?" Wolfe was courteous, smiling.

"Of course, Phillip." Maggie closed the door behind her, wishing she had a door to close on the chill that suddenly struck within her breast. She walked to her office. Inside, she went to the windows and looked out at the dismal night. Twenty minutes later she was pacing the floor, doing battle with imaginings.

The trustees had deliberated forty minutes before there was a knock at her door. George Murphy had come to retrieve her. Wolfe was conferring with Wright as she entered the board room and took her seat.

"Sister Margaret," Wolfe said, after all were again seated, "I regret to inform you that the motion made by Sister Mary Jane was defeated. The board then decided not to renew your contract. A motion to that effect was passed. The board feels it is time for a change of leadership."

Maggie was stunned. She looked down to give privacy to the shock in her eyes.

"Naturally, we appreciate..." Wolfe droned on.

Before, in her office Maggie had tried to conquer the chill with reason. Now, for a moment, she couldn't think, and struggled for composure. She looked over at Sister Mary Jane, but her superior's eyes stared down at the table. Maggie felt starkly alone.

She turned to look at Wolfe. "I also regret," she stressed the word, "the board's decision, Phillip. In the face of such lack of confidence it is clear that a defense given after your action would be both futile and personally humiliating. I will resign, effective immediately. I will have a written confirmation of that in tomorrow's mail. Good evening,

Ladies and Gentlemen." She held her poise as she left the room. She grabbed her raincoat from her office and walked out into the rain, and walked, and walked.

The doorbell nagged Charles from his sleep. He threw trousers on over his pajamas, grabbed a pullover and stuck bare feet into loafers. He raced downstairs and opened the door. Maggie was shivering. Her hair was flat and tangled with wetness. "What the...?" Charles reached out and pulled Maggie into the rectory. "I'll get some towels. You sit here by the register." He crossed the room and turned up the thermostat, then ran to get towels.

"They fired me, Charles." Her voice was a hollow whisper.

Charles did not respond at first. He put a towel over her hair and raised her hands to it. "Rub," he said. He knelt, removed her shoes, and toweled her feet and ankles. "Who fired you? From what?"

"The Board of Trustees."

"Don't say any more yet," Charles directed quietly. He ran out of the room and returned with a pair of heavy cotton sports socks. He knelt down and pulled the socks onto her feet. "I find that hard to believe, Maggie. You've been a marvelous president. The college has grown so much. You gave it life."

Maggie stood and paced the room. "Can I have permission to cry now?" Her voice trembled.

Charles crossed to Maggie and wrapped her in a bear hug. Her sobbing came on slowly and then went out of control. Charles found himself repeating, "It'll be OK. You'll be OK."

"But I worked so hard."

195

"I know."

"And I did a lot."

"I know."

"And I really loved my job."

"I know." Charles rubbed the towel on Maggie's still wet hair.

* * * *

Easter Sunday brunch was at Eileen Fogarty's home. Bill Smith arrived with a rabbit-shaped, coconut-covered cake. He put it on the table and rejoined Eileen back in the living room. He stood at the window, and then walked about the room, his face grim.

"My goodness, Bill, will you sit down." Eileen watched his restless walk. "What's got into you today?"

"I can't sit down. I can't believe what those rotten cocksuckers have done to Sister Margaret."

"Don't you use that language in front of me, Bill Smith," she scolded, and then softened at the look of the man. "It is hard to believe, isn't it. Maggie stayed with me Thursday night and all day on Good Friday. That poor dear girl! Why would they fire her, Bill? She wasn't sure, herself."

"I don't know. I saw Phil Wolfe at the bank board meeting on Friday. All he would say was that the trustees thought it was time for a change. He's hiding the real reason, that's obvious, but I can't get to it."

"The college has been doing so well under Maggie. I can't imagine why they would do this."

"Well, they did fire her, Eileen, and I'm...well, I'm worried about Sister Margaret."

"Don't you worry about Maggie, Bill. She has spunk, that one does. Oh, that's the door bell. That'll be Josephine."

Josephine rushed in. "Happy Easter! Happy Easter! Happy Easter! He is risen! Alleluia!" She looked at Eileen, then at Bill. "My goodness, Bill, how can you look so glum on Easter?"

"Easy enough Josephine. You didn't hear about Sister Margaret?"

"Of course I heard, Bill," she responded knowingly. "I suppose it's all right to tell you. I learned about it on Holy Thursday morning. The bishop asked me to come in special for a few letters. I was taking dictation when Mr. Wright called the bishop. I must say, the bishop didn't take the news as hard as you are, Bill. Why, he was almost smiling. He said maybe the next president won't be so radical."

"Sister Margaret is no radical," Bill replied.

"Well, I think she is," Josephine asserted. "Always talking about women's rights, getting all that newspaper attention. I certainly don't agree with her. God intended women to take their direction from men. It's that simple. That's the way it's always been, and that's the way God wants it."

"Oh fiddlesticks, Josephine," Eileen said. "You don't know what it's like. Maybe you've never been treated unfairly by men. You've never been exposed to a man's world except for the chancery. I worry sometimes. That makes you so vulnerable."

"Oh, Eileen," Josephine countered, tossing her head back, "sometimes you just know it all, don't you? Well, I am not as naive or as vulnerable as you think, and I'll stick with my own opinion, thank you."

"Bill," Eileen switched direction, "why didn't the Franciscan Sisters overrule the trustees? They have the power, don't they? It's their school, isn't it?"

"It's their school all right," Bill answered. He had taken a seat and his long legs sprawled out into the room. "And they do have the power. However, I know for a fact, they're not going to overrule the trustees."

"Why not?"

"The main reason is that Sister Margaret won't let them. She's given her resignation. What's more, the board has named Sister Felicity as acting president. I think Sister Margaret and the other nuns hope that the board will select her for the job on a permanent basis. Finally, if you bring it all down to the lowest level, the trustees are the school's biggest donors. So you see, there are catch-22's even for nuns."

"What are you going to do, Bill?" Eileen asked.

Bill ran his fingers through his short-cropped hair. "I'll stick by Sister Felicity as long as she wants me. Sister Margaret asked me to do that, and I respect Sister Felicity. We'll see what the Search and Screen Committee turns up. I'm reserving my final decision until then."

"Search and Screen Committee?" Josephine sounded perplexed.

"That's the committee set up by the trustees to find Sister Margaret's replacement, Josephine. There are three members: Phil Wolfe, Stanley Wright and Sister Ruth Clarence. She's head of the Music Department, a good musician, but a bit of a fruitcake when it comes to the practical. She's close to John Winslow Alexander. You can bet he'll have her ear in the selection process."

"Isn't Dr. Alexander just the most inspiring person?" Josephine was almost breathless from the thought.

"We'd better get that brunch on the table," Eileen said. She wheeled into the kitchen.

CHAPTER TWENTY-SEVEN

Elgar's *Pomp and Circumstance* filtered through the speakers. The trustees, in academic gowns, processed down the center aisle of the auditorium. Charles Mueller, dressed in his doctoral robes, followed behind. Together with the robed and tassel-capped faculty he filed into the reserved seats, front left. The faculty were followed by the graduates who filed into front seats on the right.

Charles looked around. Not many students, he thought, as he watched several of them speckle the crowd of graduates' parents and relatives. That's no surprise. I'll bet the leak from Sister Mary Jane's office on how Maggie got the axe was deliberate. The vocal student reaction is a great tribute to Maggie. That bunch out front with their placards: *Nuns Have Rights! What Happened to Fair Play? Marvelous Magic Maggie! We Want Maggie!* I didn't even know they called her Maggie. The students sense they are losing a leader who cares for them. So does the faculty. The faculty has retreated to their offices. They'll hunker down until they see what freedom is left them. It's like watching theatre lights dim. My God, Maggie, how this place will miss you.

He turned to watch as the placard-carrying students tried to push down the center aisle, where they were stopped by security personnel. Then the protestors fanned out across the rear of the auditorium, waving their placards and chanting "We want Maggie." They quieted for the graduation ceremony, but then led the procession out of the auditorium, resuming their chant.

Late that evening, Charles sat in his rectory office and began a letter. He finished it as dawn broke.

CHAPTER TWENTY-EIGHT

"Charles, you won't believe what's happened!" Maggie's excitement bolted across the wire.

"Tell me what's happened and I'll tell you if I believe it." Charles's own excitement rose. "Where are you?"

"I'm calling from New York City."

"You sound like the Maggie before the fall. What's happened?"

"I've got a job, Charles. I'm the new president of Highmount College here in New York." Her words came in a rush. "It's a Catholic women's college, a bit larger than Padua. It's all gone so swiftly I still can't believe it."

"They must have skipped the reference checks," Charles joked.

Maggie laughed. "Speaking of references, are you ready for this? The chancery here in New York did their own check on the candidates. How they got the list nobody seems to know. Anyway, the chairman of the search committee got an unsolicited call from a chancery official, whose name I don't know. It comes out that they had checked with Bishop Sweeney. He told them he had no quarrel with my administrative abilities but that he considered me unqualified because of my feminist views. He said my views on women were on the outer fringe of the Catholic perspective, that they don't conform with the Vatican on the place of women in society. He considered me unfit to lead a college of women in the, quote, 'best Catholic tradition,' unquote."

"You still got the job?"

"It was an all-women search committee."

"That's great news, Maggie. I'm sitting here thinking how good it is to hear the lilt and life back in your voice. If I may offer one bit of advice..."

"What's that?"

"Don't turn your back on the chancery."

"I'm ahead of you there, Charles. The Board of Trustees here is not filled with episcopal cronies. Nearly every member is a woman, all of them successful, most of them wealthy. The college does not depend on any male system for support. Nor is it churchy in that ugly fawning way. I've got all that up front. So, I can be into education, not fortress building."

"Padua is already missing you." And so am I, he thought.

"I miss you, Charles, and Felicity, and some others. But, we'll stay in touch. You must come and see my beautiful Highmount College."

Thoughts of Maggie continued to surface as he worked on his homily for the coming Sunday. Padua will never be the same. I will never be the same, he thought. God, it is so difficult to find You when injustice like that rides on the backs of clerics. The phone's ring brought him back. "This is Father Mueller."

"Father Mueller, this is Josephine McGregor, Bishop Sweeney's secretary." Josephine was at her formal best.

"Yes, Josephine, how are you? We haven't talked in a coon's age. How can I help?"

"I'm fine, thank you." She was all business. "The bishop asks if it would be convenient for you to come downtown to the chancery office."

Charles pictured her upturned nose. "I will make it convenient, Josephine. When does the bishop want me to come?"

"Immediately, if you can."

"I can. I'm on my way. Save a smile for me." He pictured the drawn reaction on her face.

* * * *

Frederick Patrick remained seated at his desk as Josephine ushered Charles into the office. "Come in, Father Mueller. Take a seat." He motioned Charles to a chair.

Charles relaxed into the designated seat. "Thank you, Bishop. How can I be of help?"

The bishop stiffened. "Help is not a word to describe what has come to my attention, Father Mueller. I have just this morning received a letter from Washington. The Apostolic Delegate enclosed a copy of a letter purportedly written by you to the Holy Father. That letter is dated June 5th. Did you write it?"

"I wrote a letter to the pope a month or so ago, yes. I forget the exact day." Charles kept his relaxed posture.

"I will read this letter to you. Perhaps you will then be kind enough to explain yourself."

Dear Holy Father,

"I would point out that the proper address is *Your Holiness.* You should know that." He continued to read.

> *I write because my conscience tells me I must. I write to give you the impression of the Vatican and the Curia as I see them from my spot on the globe. I write with the naive hope that this reaches you.*
>
> *It seems from here that it is important that you be right all the time, and, that being right means being consistent with past papal opinions. And yet we have seen the spin given usury down through centuries move it from a Church-condemned evil to an accepted practice. Was that consistent? It seems that being right is more important than being honest. Are the mental gymnastics on birth control honest? Do you feel what this does to people?*
>
> *You don't have to be perfect, you know. It's OK to make mistakes. What a redeeming thing it will be for the people when a pope says, "I made a mistake," or, "My predecessors were wrong."*

The bishop paused, his face reddened with anger. "I must tell you, Father Mueller, I have never been more embarrassed in my priestly life. I can only thank God this letter was intercepted before it reached the Pontiff."

Charles spoke calmly. "I don't understand. Why would you be embarrassed? It's my letter, not yours."

"I am your bishop." Frederick Patrick returned to the letter.

> *I do not understand your attachment to things monarchical, the pretense, the royal trappings, the royal brand of power. It is visible in the court with which you surround yourself. I see it in the way you shed verbal tears for the poor, but then support their*

oppressors. It shows in how you turn love into mercy so that it is at home with your aristocracy. I see it in the way you place loyalty to you and to the institution above love of neighbor. And your example trickles down the hierarchical chain. I see so much violence done to good people because of this. Are you somehow trapped? Is there no one to rescue you?

Frederick Patrick looked up. His wide open eyes fixed on Charles. His jaw quivered. "Pretense? Pretense, Father Mueller? Just who is being pretentious here? Who do you think you are to...to preach to the Holy Father?" The bishop did not wait for a reply.

Finally, it seems you have a thirst for control. I see this in how your Curia deals with theologians in their struggle for truth. I see it when you continue to impose artificial sin for infraction of artificial rules. Control by fear? Control by threat of damnation? Would a loving parent do that? This is Christ?

You don't need to control people. You are there to show people what it is to love. If love is anything, it is a free response to an invitation. It cannot be controlled.

"I don't understand how the pontiffs have survived all these centuries without your sage advice, Father Mueller." A tremor shook the bishop's head.

I know that people love you, that many more want to love you. But, people don't need you, you know. They can be loving human beings without you. They can have a deep spirituality without you. They can look at the cross and see the face and depth of God's love without you. If there is one thing you can be for your people, it is a living symbol of God's love. I suggest you get rid of those rules that have no relevance to life or to love. Get rid of those royal trappings that betray love by converting it to mercy and charity. You may come to know those who love you from those who use you.

Should you wonder who I am, I am your brother priest.

Respectfully,

Charles Mueller

The ticking of the bishop's desk clock disturbed the silence.

"Tell me, Father Mueller," the bishop's voice had a chill that matched his eyes. "Do you fancy yourself some sort of prophet? Has Almighty God designated you to speak for Him. Do you feel called to vilify His vicar on earth?"

"We are all called to be prophets, Bishop, to speak our truth openly. And that letter does not vilify."

"Really? I wonder why I don't have priests in here every day to explain their epistles to the pope. And, if you're a prophet, it's a very minor prophet indeed, Father Mueller," the bishop scoffed. "Your comments have all the weight and permanence of a...a bird's tweet, an uninvited burp."

"Perhaps you're right, Bishop. But, even if it's only a burp, I will burp."

"Perhaps you're right, Bishop. But, even if it's only a burp, I will burp."

"You need a reminder, it seems to me, that you are a priest. And you should realize that the priesthood places a greater demand on you for loyalty to the hierarchy and the Holy See."

"I cannot be more loyal than to speak the truth as I see it. Does the Holy See want its priests to be robots?"

"I will ask the questions, Father Mueller. You should thank God for the rights and privileges and status given you because you are a priest."

"No, Bishop, those things don't come to me because I'm a priest. If they did, they would follow the priests who left the ministry but are still priests. Those things come to me because I'm a cleric, a man of the institution. And I can't thank God for a status that makes me feel dirty."

"Dirty? What do you mean, dirty? There's no greater vocation in the world."

"Dirty because those privileges are paid for by a self-serving manipulation of the gospel."

"That's tripe."

"No, not tripe. The people could starve for the sacraments before the Vatican would permit a married clergy, or women priests. And why? Because a celibate male clergy protects the clerical institution. And what does that institution protect? The privileges of the clergy, especially the hierarchy, most especially the Vatican Curia who buy their job security at the people's expense."

"If you're so disaffected with the Church, why did you become a priest? Why do you stay?"

"I became a priest because of the gospel ideals. I was naive. It did not dawn on me that the institution would not reflect and support those ideals. It doesn't. I stay because I can be a priest to the people. And I just might be able to help change things. Perhaps I'll even convince you someday that putting the institution ahead of the people does violence to people in the name of the gospel."

"I wouldn't hold my breath if I were you, Father Mueller. I don't see any violence." The bishop's fingers worked his pectoral cross. "I've had enough of this discussion. You may leave. And, you will be hearing from me again on this matter."

Charles stood. "I'm not leaving, Bishop, until you hear me out. Wasn't it violence you worked on Monsignor Oberkirche when you sent him that letter of transfer?"

"You are indeed naive. How can you call a simple transfer violence?"

"Maybe I am naive, but you should not have transferred him. He had been promised he could stay there. Does it ever strike you as strange that your churchmen think more of useless specks of dogma that they do of keeping their own word. They keep the dogma down the centuries with no particular benefit to people, while they can break their word, and the word of their predecessors, with impunity. You could have asked the guys who worked in the chancery when the promise was made. They would have verified it. And Frosty might be alive today. So why did you do it?"

Bishop Sweeney stood and leaned forward over his desk toward Charles. "I did it for the Church, if you must know. It was a sound business decision. I did it because I could get maybe fifteen more years of priestly work from Monsignor Haggett vs. how many from a 79-year-old Oberkirche?"

"That's exactly my point. Everybody knows about Haggett's threat to resign if he didn't get Frosty's parish. You bought him off and Frosty paid for it." Charles paused and moved to the side of the desk. "Why did you get Tony Corsalini fired? Or do you think getting someone fired isn't an act of violence?"

"How do you know about...?" Frederick Patrick moved from behind his desk and stood facing Charles. "I don't need to defend my actions to you, Father Mueller. I also don't need to hide my actions from you. Corsalini was an embarrassment to the Church."

"What church? Your clerical church?" Charles's voice grew brittle. "That's incredible. Did you think of him also as a father? As a husband, with a family to provide for? As a human being?" He stared at the bishop. "What was your reason for sicking your mercenaries on Sister Margaret McDonough?"

"I'm getting tired of trying to instruct you, Father Mueller. What do you know about leadership?"

"Where are you trying to lead us, Bishop? What happened to the gospel? What happened to the priorities set by Christ? Love of neighbor? Do good--even to those who oppose you, who are different? Isn't that where you're supposed to lead us? How does that square with what you did to Frosty, to Tony Corsalini, to Sister Margaret, and, to God knows how many others?"

"What I did, I did for the good of the Church."

"Not the Church, Bishop, not the people. You did it for the institution, for yourself. And violence rides the underside of that mistaken priority."

Frederick Patrick's face reddened. He spoke coldly. "You are correct in one thing, Father Mueller. I am concerned about the well-being of the institution. I fail to see where that means I am not concerned about people. And right now you are my immediate concern. I will not have you spewing disloyalty and dissent in my See city. You can expect a transfer shortly to a rural parish. Hopefully, in your new surroundings you will lose your concern about violence from your bishop. And I'll make you a promise. I promise to leave you there as long as I am bishop here. And I will work with the personnel committee to select a place where the work-load will allow you greater time for reflection. I suggest you meditate on humility, Father Mueller. Perhaps then you will not feel so dirtied by your clerical status. Now get out!"

Charles turned and walked out of the office.

CHAPTER TWENTY-NINE

Bill Smith wet his forefinger with saliva and tested the skillet's heat. It was his kitchen. Eileen Fogarty had wheeled up to the table. "I'll wait for Josephine before putting the eggs on," he said. He turned to Eileen, his gray eyes showing resignation and sadness. "Have you heard the news about Padua?"

"What news, Bill?"

"Phil Wolfe called me last night about a banking matter and filled me in, so it's no secret. Once again man rises to his level of incompetence. They are replacing Sister Margaret with Monsignor Nick Deutschman. The Peter Principle gets confirmed once again. And that's not all. Phil didn't say so, but it's obvious to anyone who knows Deutschman. Because Deutschman's about as bone lazy as they come, he will restructure the administration. The board has already given its approval as part of his employment contract. Are you ready for this? John Winslow Alexander comes in under the new title of executive vice president. He's Deutschman's choice, no faculty consultation, nothing. Can you believe that lazy ass, Alexander, coming up smelling executive?"

"Ouch! That is a bit hard to picture."

"Wait, there's more. Alexander gets an assistant, an assistant vice president. Can you believe it?" Bill's eyes blazed. "All that means is Deutschman gets more time on the golf course, if that's possible, and Alexander gets more time to flit about with the upper crust. And Padua goes to hell. I pity that poor damn assistant. He or she will carry the load, and carry the blame when the shit hits the fan, but never hear a word of praise. Eileen, there ought to be a goddam law against such insanity." He sat down, drained.

"There's no need for goddams, Bill. You just stop it now." She was about to say more when Josephine came rushing in.

"Did you hear? Did you hear? Did you hear the terrible news?" Josephine's high voice broached the terrible news, breathless and grating.

"I've been listening to Bill's terrible news, Josephine. What's yours?" Eileen asked.

"They're taking our wonderful Monsignor Deutschman from St. Martha's and transferring him to Padua College as the new president."

"You should keep him at St. Martha's," Bill sniped. "Why don't you ask Bishop Sweeney to let you keep Deutschman at St. Martha's?" he repeated. "You're close with the bishop, Josephine. Tell him you'll quit otherwise. That'll scare him."

The sarcasm didn't connect. "Don't be silly, Bill. I couldn't do that. Besides this is a plum in Monsignor's hat. He'll be a bishop someday, you can bet on that."

Bill kept to his own track. "I believe you, Josephine, even though the Peter Principle doesn't need further verification. It's bad for you that he leaves. It's bad for Padua that he comes."

"I don't know what I'll do if they bring in the wrong kind of priest at St. Martha's," Josephine said.

"Wrong kind of priest, Josephine? What do you mean?" Eileen knew.

"You know, the kind that throws out all of our wonderful statues and devotions, and brings in all that modern stuff. You, know, someone like that Father Mueller. It won't be

214

him for sure. The bishop is packing him off to the country," Josephine said snippily

"I happen to be very fond of Father Mueller," Eileen countered. "What do you mean, packing him off?"

"It's not out yet, but he's being transferred."

"Oh my! I hate to hear that. I must finish his painting. When is he going?"

"It's not set yet. In a few weeks, I think."

"I shouldn't care, I'm out of there," a preoccupied Bill said.

"What do you mean, Bill?" Eileen's concern showed.

"I quit. I gave my resigntion to Sister Felicity yesterday. It's effective on her last day as acting president, which will be the day before Deutschman and Alexander take over."

"You're going to miss Padua, aren't you?" Eileen replied.

"I'll find something else to keep me busy," Bill responded without conviction. "I feel bad for Sister Felicity. She's a brave and competent woman but when she heard that Deutschman got the job she got physically sick, vomited right there on her desk."

"The poor dear!" Eileen's eyes grew sad.

"The trustees are planning a big inauguration ceremony for Monsignor Deutschman and Professor Alexander in October," Josephine chimed in. "I'm sure they'll have a big party."

CHAPTER THIRTY

Monsignor Michael Higgins sat in the bishop's office. His dark eyes peered from a flushed face at His Excellency, Frederick Patrick Sweeney. "Your friends are correct, Bishop. If you recall, I have suggested the same change several times over the past years. Josephine simply does not present the image that ah...serves your best interest."

"I must admit, I've had the same feeling at times, Mike, but she has been such a competent secretary." He hesitated, knuckle stroking his jaw. "But you agree with Wright and the others?"

"Bishop, her appearance alone is enough to rule her out. I don't want to sound uncharitabale but look at her, a nineteenth-century hairstyle, that prissy bun, her dyed harlot-gold hair." Higgins pinched his words. "Look at her prune face, her prudish manners, those squinty eyes behind thick glasses. And look, no, listen to her high-pitched grating voice. God bless the woman, it's not her fault, and she has many good qualities, but there is no way that she can present the proper, professional comportment that is essential to one who represents you so closely. Think of the countless daily encounters she has in your name and as your intermediary. This is not a personal thing, Bishop. We're dealing here with the office of the bishop." Higgins stopped and waited.

"What do you suggest?"

"That we transfer her back to the Marriage Tribunal. Janet Green's departure has left a vacancy there."

"Don't you think that might seem like a demotion and hurt her terribly?"

"Perhaps, Bishop, but we can't allow hurt feelings to deter us from what is best for the office of the bishop."

Frederick Patrick was thoughtful. His knuckle continued its massage. "Perhaps we can put her transfer in the context of the new revision of the Code of Canon Law. That's effective next month and it will be an event of considerable magnitude if you consider that the last change was in 1918. Lots of chancery titles will change. Do you think we might put her transfer into that larger picture, soften the blow?"

Higgins screwed his face into a frown. "It's not a good fit, Bishop. Those changes go only to members of the diocesan curia. She is only a secretary." He thought a moment. "I suggest we put her transfer in the context of a cross-training policy. Our new policy will be that all secretaries must be able to handle any job in the chancery. That means they have to move from time to time. I'll move Josephine out and Mary Dunne in. We can soften the blow that way. As time goes by we'll simply let the situation stabilize. Mary's dictation skills aren't up to Josephine's, but she'll learn fast enough. And she does carry the right image."

"You handle it as you think best, Mike. Is there anything else?"

"No, Bishop. I'll get at this transfer right away."

Shortly afterwards, Josephine walked out of Higgin's office. Her clenched jaw held back a cry of pain. Her eyes released it. She walked shakily to the door and left the chancery.

An hour later she was still on her knees in front of the huge crucifix at St. Martha's. You want me to suffer, don't You Lord? she prayed. You want me to feel Your hurt and Your humiliation. Well, I feel it. Tears ran freely, eroding her carefully placed makeup. The bishop could have kept me if he wanted, but...She banished the thought. He's so good. And his problems are so immense. How can I expect him to think of my feelings? I know You want me to feel special, Lord, because You give me sufferings just like Your own. But it hurts so much...

* * * *

Together for Sunday brunch, Eileen Fogarty looked at Bill Smith. "Josephine will be here any minute, Bill. She's hurting. She needs our support."

"What a lousy break," Bill responded. "I don't agree with ninety per cent of Josephine's religious ideas, but where it really counts, she's as good as silver. I feel so sorry for her. Did she say why they transferred her?"

"They, Monsignor Higgins, that is, told her they were implementing a new policy of cross training. All secretaries must be able to do all jobs, that sort of thing."

"That's baloney. It's clear that Bishop Sweeney wanted her out and used that for cover. She'd still be his secretary if he wanted her."

"I hate to think that..."

"Eileen, you're too goddam kind. So is Josephine, usually. You both refuse to believe that the clergy can operate with less than noble motives. And you know Josephine as well as I do. I'll bet she's knee-deep in martyrdom by now. It's God's will and all that. That's bull. We've got a bishop who doesn't give a shit about people--except in the abstract, of course. He loves mankind to death, but not this

219

insignificant person, Josephine. It's all crap." Bill's gray eyes smoldered.

"Calm down, Bill. Let's focus on our friend."

"Hello, Hello, Hello, dear friends." Josephine walked in the kitchen door.

"How's Mill Valley's best secretary?" Bill greeted. "Did I ever tell you, Josephine, about all the complimentary things said about you by the business community in this city?"

"No, Bill, but thank you for those kind words. I can use some nice words about now."

"Are you going to the inauguration this afternoon?" Eileen inquired.

"No, I'm not," Josephine responded. "That sort of thing is for more important people."

CHAPTER THIRTY-ONE

"Good afternoon, Sister Felicity."

Felicity's face brightened. She turned her large body to face Charles. "Well, a good Sunday afternoon to you, Father Mueller. It's so good to see you. We miss you here at Padua. Why don't you come back?"

"I miss all of you too, Sister. And I'd like to be here. Unfortunately, it's just too far to commute." Charles didn't mention that Nick Deutschman had let him know, as only Nick can, that there was no longer a position for him at Padua. Nick intended to pull that school together, restore old-time Catholic tradition, and instill solid traditional theology. Charles simply would not fit in the Padua of Nick's vision. And Nick had already tightened the screws on the college. Because of the student demonstration at the June graduation he had banned all such demonstrations under penalty of expulsion. The students had pulled back. The faculty were still bunkered in their offices.

"It was good of you to come for the inauguration." Felicity's statement came out like a question.

"Actually, I'm killing two birds with one stone, Sister. I'm in town to pick up a painting from Eileen Fogarty. Since she is just across the street I decided not to miss ole Nick's elevation. All of the clergy received invitations."

They were gathered in the Administration Building, the board of trustees, members of the faculty and administration, and the clergy, waiting to process down the street to the auditorium.

"What are your plans, Sister?" Charles knew of Felicity's resignation.

"I'm not sure yet, Father. I have a sabbatical coming and I'm using it now. Sister Margaret has invited me to join her at Highmount. I'll go there to see her and review her offer. But, I haven't decided anything yet. Say a prayer for me."

"I'll do that, and I wish you the best. You deserve it." Charles keenly felt sadness at Padua's loss of this great woman. Thay have lost Maggie, and now, Felicity, he thought. Students and faculty trusted them both.

At that moment, the faculty member master of ceremonies rapped on a table for attention. "Ladies and gentlemen, we will begin our procession now. The trustees will lead. They will be followed by the faculty, who will be followed in turn by Vice President Alexander and President Deutschman. Members of the clergy will come next, with His Excellency at the end of the procession."

October's brightest blue weather greeted them as they emerged from the building. As they filed two by two down the street they were also greeted by a small contingent of placard-bearing students. These students, most of whom were fanned out at the auditorium's entrance, were silent. Their clear act of disobedience was supported by dozens of other students standing about. The placards read: *Gays are Human! Christianity is About Acceptance! Christianity Without Questions? Lesbians Have Rights Too!*

Stanley Wright's mouth dropped when he recognized his eldest daughter, Sandra, at the auditorium door. She stood, eyes drilling his own. Her placard read: *Lesbian Love/ Christian Love.* She held the hand of another young woman who helped her hold the placard.

Charles observed it all. These kids are saying we are who we are, he thought. Accept us. Wow, Maggie, if you were only here to see this.

222

The procession moved unhindered into the auditorium. When the placard bearers tried to follow they were stopped by security personnel.

As the speeches droned on and on, Charles scanned the faces of the trustees as they sat on the stage. He saw the shock that remained on Stanley Wright's face. Sister Mary Jane Plummer and Sister Felicity sat next to each other. Both women wore smiles that seemed to hold in laughter. Charles knew their thoughts. What goes around, comes around.

I wonder if Stanley will insist on Sandra's expulsion? he mused as he left the auditorium. No, the board will extemporize all the necessary denials, place the blame on Maggie's permissiveness, and drop it. Sandra's daddy will save the whole bunch in order to save the one.

Charles walked down the street to Eileen Fogarty's.

Eileen had the Holy Spirit painting on her easel. "You take it and look at it awhile," she said. "I'll want it back in a few months for any changes you want, and for a few final touches of my own."

"It's beautiful, Eileen. I know exactly where it will hang."

*   *   *   *

Angela Pannetto told herself she'd died and gone to heaven. Joy moistened her eyes as she surveyed her quarter acre, fresh now from the traces of the plow and disk harrow. The farmer had even returned with his spreader and layered it with manure before harrowing it. He will harrow again in the spring. My garden, she thought, at last, my very own garden, and a big garden. Born into a truck farming family she had

223

missed the soil of her youth during all her years as housekeeper, two decades for the dear old Monsignor Mazzola, and nearly five years now for Father Mueller. The parishes had all been urban till now. Four tomato plants along the rectory wall in Mill Valley had given scant relief to her farming impulse. She lay awake nights now, her mind designing and redesigning her plot of ground: ordered rows of tomatoes here (from now on my spagetti sauces will have only garden-ripened tomatoes), beans, peppers, squash, there and there and there; raspberry canes, strawberries, rows, rows, rows. Seed catalogs arrived daily in the mail.

Charles watched from his study window. He had worried that Angela's attachment to St. Stephen's would travel with her. He felt relief at the sounds of her enthusiasm for St. Patrick's. At heart Angela was country, and she had come home. Charles, himself, experienced a sense of homecoming. He watched the diminutive Angela measure spaces between rows, pull a stake, move it, and reset it. The garden space had been cut from a parcel of land designated for cemetery use in the distant future. She has garden enough there for the whole community, he thought.

CHAPTER THIRTY-TWO

It's like watching carrots wilt, Charles thought, as he looked back on their first year at St. Patrick's. Country time treads slowly in the experience, yet races in the looking back. He stood at his study window and watched Angela, busy at her harvesting. She packed tomatoes, peppers, and beans into brown paper bags and loaded the bags into a wheelbarrow. He knew the wheelbarrow was only transport to Angela's car. She had made her lists earlier in the summer, people who had the need and would welcome the gifts.

Charles returned to his desk. He checked his lists, markers from his fence walking. He had a daily and a long range list, and he checked each one every morning, revising them and checking off item by item. He looked at the long range list. *Check Parish Roster* had been there since his coming. Today he checked it off. He had now memorized the names of the one hundred forty families, and, from after-church encounters, could put names to the faces of all of them.

Item number two on the long range list: *Review School Status*. School would start in two weeks. Eighty-six kids were registered for the six grades. And the principal, Sister Marian, had that business all organized. He felt fortunate to have three nuns at the school.

Item number three on the long range list: *Apartment for Director of Religious Education.* He checked it off with a flourish. Nick Deutschman's loss had been Charles's gain. Nick had hauled the theology department staff in for an inquisition early in the spring semester. He had opposed Sister Rosalie's Vatican II theology with his own Baltimore Catechism brand, and she lost. Self-esteem forced her, in language never before heard from Rosalie, to tell Nick where to stick the Baltimore Catechism and her job.

Charles had raced back to Mill Valley with a contract, and Rosalie had signed it. She was due here in ten days and her apartment had been rented. She would teach Charles how to be effective with grade school kids, he hoped. She would also bring all the books he had left at the college, and invariably forgot to pick up whenever he went back to Mill Valley.

Number one on the daily list: *Sunday homily.* Whether 'tis nobler to work on the Sunday homily or help Chester Langowski with his haying? He tossed the list on the desk. He had heard that Chester's son, his main help, had broken an arm. Chester's hay had been cut and was ready for baling. It's only Wednesday, Charles thought. He'd help Angela with the vegetable deliveries and then go haying. Like Angela, he felt the tug of the soil.

Charles felt the pain the next morning. If there's a muscle that doesn't ache I wouldn't know where to look for it, he thought. Pulling vestments overhead for morning mass pained his arms and shoulders in a way he hadn't experienced in years. One more day and my haying season is over, he thought. Lift that bale, tote that bale, stack that bale. It hurts so much this better be good for me.

Friday's list: *Do Homily!* There won't be time on Saturday. The homily finished, he went into his bedroom and removed the Holy Spirit painting from the wall. Back in his study he wrapped it in cardboard, put it into a carton, and taped the carton shut. He addressed the carton to Eileen Fogarty. She wanted to finish it "before we're both dead and gone" she had written. A half hour later he carried the carton out to the road to meet the mailman. "How's it going, Jack? Your timing's as good as always."

"It's goin' good, Father. I take pride in my timing." Jack lifted the visor of his cap and peered up from the car window.

"Looks like sunny days for the weekend."

"That's the way I hear it, too, Father. Here's your stuff. You want me to take that box?"

"I'd appreciate it, Jack." Charles opened the back door and deposited the carton. He handed a five-dollar bill to Jack. "Will that cover it, Jack?"

"Sure ought to. I'll square with you tomorrow."

Walking back to the rectory, Charles flipped through the mail. There was a letter from Tony Corsalini in the stack. He ripped it open. Tony and Kate would be back visiting family in early October. "How about three days of muskie fishing?" Tony wrote. Back in his study, Charles penned a quick reponse. The phone rang as he slapped a stamp on the envelope. It was Maggie.

"I'm flying back to the motherhouse for a meeting. The meeting is on Sunday. Can we get together next week? If it's OK, I'll drive out to your place. I want to get the sight and feel of where you live so I can picture you better." No soil could tug at him like the sound of her voice.

Charles scratched Tuesday's list and wrote MAGGIE in bold letters.

Saturday's list: *1. Attend Festival. 2. Evening Mass.*
September dawned today and the people held a harvest festival in Posnan every year on the Labor Day weekend. The small village hugged the flat sandy soil like fishing huts on a dull frozen lake. The church tower and spire alone rose above the single-story buildings. On festival day, flags and banners bloomed everywhere and brought color and luster to the drab main street.

Today was also the festival parade day. Long enough to stretch through a dozen Posnans, the parade brought out the entire village, and brought in all of the farm families connected to Posnan through St. Patrick's parish. The people lined the length of Main street, County trunk C, several deep. They had lawn chairs set and reserved along the parade route hours before the 1:00 p.m. starting time. The parade was assembled in a pasture on the town's west side. It would follow a route that went the three-block length of the business district: three bars, a cafe, hardware store, garage, two gas stations, and half a dozen residences. At the end of the route, the parade would circle onto a back lane that returned it to its assembly place.

Lights blazed and sirens screamed from the John's Point sheriff squad cars. They led the parade. They were followed by a mix of displays. There were the town fire trucks from the 1940s and the horse-drawn floats, farm wagons decorated with colored paper and streamers and drawn by groomed Percherons and Clydesdales. The floats carried members of AARP, the Lion's club, a senior accordion band, the 4-H club, the church choir, small polka bands, the American Legion, and its Women's Auxiliary. Angela, dressed pioneer style in polka-dot calico, rode in a Conestoga prairie schooner. The school band marched, as did a military band from nearby Fort McBee, along with every other group that could commandeer a wagon. Riders in western attire rode scrubbed horses. Clowns walked the sides of the road or dwarfed the miniature cars they drove. Riders tossed candy from the parade vehicles as they passed. Kids dodged paraders to seize their share. The knights of Columbus from St. Patrick's collected nearly a third of the school's annual operating expense from a brat and beer tent set on the parish lawn.

Charles walked the parade's route along the sides, greeting people as he went. The undisguised affection that greeted him left him feeling embarrassed. The parade lasted just over an hour, but the celebration continued. He bought a beer from the K.C.s. The odor of sanctity at the five o'clock mass would smell like beer. He watched as present thirsts were quenched and future ones averted. This is good, he thought. It's good to see and it's good for the people. We don't celebrate enough.

Charles returned to the rectory at four o'clock to change and prepare for the liturgy. He was changed and in the study when he heard Angela answer the doorbell and escort someone to the office just off the entrance. She appeared at his door to announce, "A young man, Tommy Krakowski, would like to see you. Do you have time?"

Charles looked at his watch. "Sure, I'll make time." He knew the Krakowskis were parish members, knew them by sight, but had had small chance for conversation with them. He opened the office door to find the young man standing and facing him. High school senior? College? Charles reached out to shake hands. "You're Tommy Krakowski? I'm Father Mueller. What can I do for you today?"

Charles's attention was trapped by the turbulence in the boy's eyes. He did not see the gun in the hand raised to meet his own. A shot, a searing in his chest, a "wha..." followed by a gurgle. He did not hear or feel the four additional bullets as they entered his body.

CHAPTER THIRTY-THREE

Angela reached the office in time to see the gun raised, fired, and a small red hole appear in the boy's temple. She raced out the front door to the men at the beer tent. As they stared at her, uncomprehending, she pointed to the house, formed soundless words, and collapsed.

The dazed congregation milled about the rectory lawn and watched the flashing lights of sheriff squad cars and paramedic vehicles approach across the flat prairie. Inside the rectory several men kept helpless watch. Outside, men had stopped refilling their beers. Women had rosaries in their hands. Parish women were attending to Angela. She lay stretched out on a picnic tablecloth. When the sheriff's deputies and paramedics arrived, deputies led the procession into the rectory.

The paramedics emerged shortly with Charles on a stretcher. One beefy paramedic thumped Charles's chest in CPR rhythm as they walked to the ambulance. Another held an oxygen mask to Charles's mouth. They were quickly on their way to the hospital in John's Point. Minutes later a second stretcher was carried out by other paramedics, who were also quickly on their way.

Father Charles Mueller and Tommy Krakowski were pronounced dead upon arrival. The paramedics relayed the word back to the deputies in Posnan and they, in turn, broke it to the people who stood waiting in front of the rectory. The news spread quickly. The deputies resumed their work. That it had been both a killing and a suicide was evident in the bloody office. Angela's testimony, even through shock and near incoherency, verified the circumstantial evidence. The people drifted aimlessly, unable to tear themselves from the parish grounds. Many wept. The demands of cattle, milking that couldn't be delayed, eventually pulled the farmers away. Townsfolk stayed until darkness and then wandered off.

* * * *

Maggie got the news when she arrived at the motherhouse in Mill Valley that evening. Sister Felicity intercepted her at the door.

"What's up, Felicity? You look so down. It's not like you," Maggie said as Felicity led her into a small reception room.

"I have terrible news, Sister Margaret," Felicity said, fixing Maggie with doleful eyes. "Father Mueller has been shot. He's dead."

Maggie stared at her friend in shock and disbelief. "No! It's not true," she cried.

Felicity stood there. Her eyes began to tear.

Maggie read the truth in Felicity's face. "How? When?" Her words came out with stunned dullness.

"This afternoon," Felicity answered. "Someone called the college chaplain. I only learned of it minutes ago." Felicity saw the pain and shock in Maggie's face and opened her arms.

Maggie fell into her friend's warmth and the tears came. She sobbed on Felicity's broad shoulder until numbness blocked the flow.

Maggie's feeling of loss intensified later in the privacy of her room. Tears flowed again, uncontrollably. She felt alone, anguished and empty. "I never felt alone as long as you were somewhere, Charles," she said aloud.

Of a sudden her agony pushed its conclusion up through the layers of denial. "Oh, God! Why didn't I see it? Why didn't I admit it? Why didn't I say it to you?" She looked up and shouted at the ceiling, "I love you, Charles Mueller!"

* * * *

Frederick Patrick Sweeney was entertaining friends when the news reached him that same evening. The predinner cocktail hour had progressed far enough to mellow the group. He was pried away by Father John Curley with the message that Father James Brennan from St. Matthew's in John's Point was on the phone. "He says it's a matter of great urgency."

The bishop suppressed annoyance at the interruption. "This is Bishop Sweeney, Father Brennan."

"Bishop, I have the sad duty to inform you that only a short time ago Father Charles Mueller was shot and killed at his rectory in Posnan."

Frederick Patrick felt a momentary disorientation. "Would you repeat, Father. I'm not sure I understood."

Brennan repeated.

"How did it happen?" asked the bishop, struggling for perspective.

"I don't know all the details yet, Bishop. It seems that he was shot by a boy just out of high school, a young man named Thomas Krakowski. Krakowski then shot himself and is also

233

dead. That's all I've learned so far. As I learn more I will keep you informed."

"Please do, Father Brennan." A disturbed bishop returned to his guests.

"I hope it wasn't bad news, Your Excellency." Phillip Wolfe caught the bishop back into a male cluster.

"It is terrible news, Phillip. Father Charles Mueller has just been shot to death in Posnan. I don't have many details yet." He absently reached for a fresh glass of wine from a passing tray. "I will never understand what provokes such violence." He stopped abruptly, his eyes without exterior focus. Krakowski, he thought, Thomas Krakowski. His reliable memory prompted a replay of his meeting with the Krakowski parents. "Our son, Tommy," they had said.

"Are you all right, Your Excellency?" a concerned Phillip Wolfe inquired.

CHAPTER THIRTY-FOUR

Bishop Sweeney sat in the dead Charles Mueller's study and reread Mueller's credo a third time. He turned and looked down at the partially harvested garden. The rain had turned from drizzle to downpour and the bishop was suddenly seized with a chill. Perhaps it was this second-story view downward that triggered the warm memory....

Two years earlier, Bishop Sweeney was seated on a bandshell stage, looking downward at the activity in a small Chicago park. Pigeons swooped. They landed briefly beside moving feet to grab a morsel, and then took wing again. "There's many a Mick alive today because of the pigeons," his father had once told him. "Manna from heaven they were in the depression days."

Frederick Patrick's mother, Hilda, and his sister, Mary Colleen, sat to his left. The two women gabbed while Mary kept a warning eye on four imps who were seated in the front row off-stage. A six-member brass band occupied the opposite side of the band shell. They tuned their horns in subdued toots. Crowds milled around the small park where green space was just large enough to hold a junior league baseball field and several horseshoe pits. Men and women filed into the park from the neighboring bars.

The crowd quieted at the sound of approaching sirens. Bells began to peal from St. Bridget's imposing steeple. The church stood straight down the green and across the street from the band shell. The blast of sirens blended with the spinning lights of squad cars as a caravan pulled to a stop between the church and the band shell.

A uniformed police officer left the driver's seat of a slick black limo and opened the rear door. First to exit was His

Honor, Mayor Mulligan. He turned on his toothy grin and waved to the cheering crowd. Next to exit was Alderman Patrick Daniel Sweeney, a white-maned, tall, thin man, whose eyes carried surplus laughter from his lips. The crowd roared and Alderman Sweeney waved.

The Sweeney family on stage and the assembled crowd were here to honor Patrick, their retiring alderman. This park would be renamed for him. The honor that had gone to a now forgotten civil war general, who still rode his horse in the corner of the park, was being transferred. It now belonged to Alderman Patrick Daniel Sweeney, at least for a while. "Give flowers to the living," his fellow aldermen had said to the press.

Officers flanked Sweeney and the mayor as they paraded to the band shell and climbed the stairs. Elfin-faced Mickey Keegan, fellow alderman and friend of the honoree, was master of ceremonies. He stood at the podium. "My fellow Chi-caw-go-ans," Keegan began. "It is my pleasure to give you the son of Alderman Patrick Daniel Sweeney, His Excellency, Bishop Frederick Patrick Sweeney." Keegan waited for the polite applause. "Bishop Sweeney, a cut off his old man's tail, if I say so myself, will give the invocation." He nodded to the bishop. "If you please, Bishop?"

Frederick Patrick walked to the podium and adjusted the mike upwards. "Heavenly Father," he began, "the man we honor today is first of all my mother's husband and my father. He is a man who was always there for us. He is also a man who has been a neighbor to his neighbors. He was always there for them. And he's still there for all of us. We thank You for this man, Patrick Daniel Sweeney, and for the good he brought to our lives. May this park, which Chicago and Chicagoans re-dedicate today in his honor, be a reminder to us of his example. Amen."

"Amen," the crowd thundered as the bishop returned to his seat.

The chill of Mueller's study urged Frederick Patrick's consciousness back to the present. The bishop faced again the stark reality of Mueller's death and felt his own need for a respite. He gave way again to the warm memory...

The mayor moved his short plump frame to the podium and flashed his newsprint smile. "He was *always* there for you," he started, picking up on the bishop's theme. "During the depression when him and me was only hacks workin' our way up the ladder, Paddie Sweeney was *always* there for a fella. How many brown sacks full of groceries turned up on your porches when you were out of work, when times was tough? Huh? The mayor paused and monitored the crowd for nods of agreement. "He was *always* there when you needed somethin', a job, a friendly word to a judge or the police, bail money, gettin' the potholes fixed, when you needed a fin or a sawbuck. Huh? You trusted him so you made him your alderman. He's been that now for thirty years, and for all that time he was *always* there for you..."

Bishop Sweeney felt the cold present flutter back. He closed his eyes and shut it out again...

The mayor's talk turned to his great city and his own great record. Frederick Patrick's eyes wandered among the people, looking for members of his old neighborhood crowd. Slip Knot O'Reilly leaned against the *Fresh Baked Pretzels and Hot Cinnamon Rolls* wagon. Slip Knot, never himself cornered, had someone between himself and a tree. They would be talking insurance, Slip Knot's bread and butter.

The totally uninhibited Bats Brogan leaned against the backstop for home plate. Tomorrow, when the old crowd gathered for lunch at Kavanaughs, Bats would greet him with the same line that went back to Frederick's ordination day. "Are ya gettin' any, Ferdie?" Then Bats would have a fit of unbridled laughter.

Straight down the center path, Beagle O'Leary stood in dapper elegance. Beagle had wanted to be a lawyer from the first grade. Now his name was in the title of a national status law firm, his office at the pinnacle of a loop skyscraper. They will all be there tomorrow and I will be 'Ferd' again. On bad days it had been Nerd Ferd or the more formal, Ferd da Turd. Men who knew you as a boy don't let you forget. The thought was warming.

When the mayor finished his talk he took the back stairs down to the ground. Out of the crowd's sight he walked to his limo, which had moved to the rear of the bandshell.

Keegan's voice pitched higher with enthusiasm. "My fellow Chi-caw-go-ans. The man of the hour! The man immortalized in this park! The man hisself! I give you Alderman Patrick Daniel Sweeney."

The elder Sweeney faced the crowd and grinned while they hooted and whistled. When they finally hushed he thanked them for the great honor, but gave full credit to his wife, "who deservedly is the one most entitled to this honor. And now, I want to shake the hand of each and every last one of you." He walked down the stairs and began to work the crowd as if an election were in the offing.

The bishop joined his father in the crowd. The elder Sweeney knew all these people by name. He knew their histories and he knew their children. Frederick Patrick realized that he had inherited that same political asset, the genes of his father's memory. He found that he knew their names also, from long ago walks through the ward with his father.

238

"Paddie, I voted for ya every election since ya started running. And so did my dad, and his dad too." Donald Dugan grabbed his turn with the honoree and put his grinning face, breathing whiskey, into that of the alderman.

Paddie put his arm on Dugan's shoulder. "You've been a friend, I'll say that for you, Donald. And I appreciate your dad and grandad trudging up from their graves to vote for me, like they did all these years. Now, you tell McNally to set one up for you. He's running a tab for me today. There ought to be more like you, Donald." He gave Dugan a pat on the back and Dugan turned toward McNally's Pub.

After an hour the pair stopped at a pushcart and Paddie ordered two hot dogs. "Catsup, mustard, and lots of relish, Arnie. How's business been?"

"Been good, Paddie." The heavy mustache bounced with the words. "The cops been layin' off me. Thanks for that." Arnie handed a hot dog to each man.

"Pushcarts are illegal in Chicago, now," Paddie explained to his son. "But guys like Arnie have to make a living."

The bishop nodded his understanding. "Did you really get the votes of dead men?" he asked, recalling Dugan.

Paddie took a huge bite of the hot dog. Catsup and mustard ran from the sides of his mouth. He swabbed it with the napkin that had served as a sanitary grip for the hot dog. "Absolutely," he replied. "Now, I never held a rally in a cemetery as far as I can remember, but I was aware the votes were coming my way."

"Many dead men?"

"Only if it was a tight election. I needed the dead ones when I got started. The ward was all Irish and Germans then, but so was my opposition. Once the people got to know me,

239

and I got to know them, I needed fewer graveyard votes. So, by the time other groups filtered into the ward, some Italians, Poles, Hispanics, and Blacks, I was a shoo-in. Some of the guys, like Donald Dugan there, just kept on doing it. You think it was wrong what I did?"

The bishop took a careful bite of his hot dog. There was a moment of discomfort.

Paddie felt it and continued. "Let me tell you something about helping people, Son. Up there on the stage a while ago you called it 'being there for them'." Paddie waited for eye contact with his son. "You gotta get elected first. If you don't have the power, you don't have zilch for helping people." Paddie stuffed the remainder of the hot dog into his mouth and chewed it down to where he could talk.

"There's two things about the power I got," he continued calmly. "How I got it, and maybe my way wasn't perfect, and how I used it, and there I don't apologize to nobody."

Frederick Patrick took another small bite of the hot dog and dropped the remainder into a waste can. "I don't...," he started, but was interrupted by his father.

"The people gave me my power. They kept me in power. And you can see how they're with me now. There's another kind of power, Son, your kind. You weren't elected by the people. You got it from the pope and his gang in the Vatican who claim their power comes directly from God. You only had to politic those elite guys and they made you a bishop. You're a grown man, Frederick Patrick, and you're probably a better politician than I am. I never politicked at that level. So you sure as hell don't need any advice from me. But, think about this sometime. In my heart I think God works it my way. If the people don't give you power, you really don't have power, pope, bishop or not."

Both men let their eyes sweep the park. Paddie continued. "There's another lesson I've learned. Listen to the people. When winds change, have the courage to change yourself. For me, it wasn't just to get elected. I knew the wisdom of my constituents. Why is it the Church leaders never listen to the people? Do they think they can't find the Holy Spirit's voice in the peoples' experience? Take my word. They'll listen someday. When the wind blows in that direction, Son, fly with it..."

The bishop heard the front door open downstairs. Father John Curley's voice called, "Bishop?" It was time to go home.

* * * *

Sleep eluded Frederick Patrick that night. When it came it was fitful, hazy, dream-filled: a young boy, a gun, shots, a surreal farm couple, the smell of the barn, a blood drenched roman collar, bizarre acts of violence to nameless people, grotesque wounds. Charles Mueller's face floated in and out. In one scene Mueller called out, "Do you understand now? Violence! Violence! Violence!"

The bishop woke perspiring. An understanding fastened one tentacle after another to his mind until there was no escaping its grasp. He got out of bed and pulled on his robe. In his study, he paced, remembering detail after detail. Wide-eyed yet at dawn, he knew that Charles Mueller had changed him forever.

CHAPTER THIRTY-FIVE

On Monday, Labor Day, outside of Posnan, a graying deputy sheriff knocked on the farmhouse door. The stout red-eyed woman who opened the door stared, wordless, at the officer.

"Mrs. Krakowski?"

"Yah, I'm Mrs. Krakowski. What is it?" Her words were flat.

"I'm deputy John Pritzlaff, Mrs. Krakowski. I'm an investigator with the sheriff's department. I'm sorry about the tragedy with your son, but, there are a few matters that need clearing up. May I come in and talk with you and your husband?"

The kindly voice coming from a uniformed authority figure attached to a need in the woman. Since Saturday, she and her husband had suffered alone. Neighbors, from fear or confusion, had avoided them and provided no comfort. "Come in, please. Stanislaus," she called to the next room, "there is someone here to see us."

Detective Pritzlaff spent time building a shelter of trust. They should not shoulder any blame for the tragedy. Everyone knows they are good people. He was only seeking to discover what had motivated the boy. Perhaps they could help?

Mrs. Krakowski was the first to break through her wall of fear. The bishop was wrong. God did not take care of my boy, she told herself. "The boy was abused by the priest."

"Sexually abused?"

"Yah, it was sex."

"Father Mueller sexually abused the boy?"

"No, not Mueller. It was two priests ago. His name was Crocker. The boy was never the same. He kept getting worse. Always troubled, always getting into trouble."

"Did you report it to anyone?"

"Yah, we went to the bishop."

"What did the bishop do about it?"

"He got Crocker out of here."

"What about Tommy?"

"He didn't do nothing for Tommy. He said God would take care of Tommy. We was to shut up about it."

Pritzlaff stopped probing. He knew enough. No need to scrape open all the wounds. He spent time building strength in them, performing that side of police work seldom given publicity. "You are good people. All your neighbors say that. I know how hard this is for you. I will pray for you and for Tommy."

Later that same day a second car pulled up at the Krakowski farm. "Wait for me here, Father Curley." Bishop Sweeney walked to the house, knocked and was admitted. He emerged an hour later.

* * * *

Word of the bishop's presence spread quickly and the church was full when the funeral column began its slow procession up the center aisle. Eyes that moved from the coffin to the bishop who followed the coffin showed bewilderment.

When the introductory rites had been completed, Bishop Sweeney stood, without pulpit, in front of the congregation.

"We are here to bury Tommy Krakowski," he began. "Three days ago Tommy shot and killed Father Charles Mueller. He then shot and killed himself. That sounds straightforward enough, doesn't it." The bishop hesitated. "I am here because their deaths were not simple, and were not straightforward."

The bishop cleared his throat and looked directly into the eyes of a number of parishioners, one after another. "We don't know what web of thoughts motivated Tommy to fire that gun. We do know when he began to be troubled and we know what caused that. You see, at one point in his life Tommy was sexually abused by a priest. That priest was not Father Mueller. You should know that." He paused again. The church was soundless.

"When Tommy's good parents made me aware at that time of the abuse, I dealt with the priest. He will never again be in an official position where children are under his care." The bishop's voice broke as contained emotions began to surface. "I failed, however...I did not attend to Tommy's needs. If I had done so, perhaps Tommy would not be here today for burial. What I did not do, and could have done, and should have done, has placed my finger on the trigger with Tommy's."

The congregation remained hushed, but now there were tears showing, on the bishop's face and in the eyes of most. "Pray for Tommy," the bishop continued, "but don't blame him. Surround his good parents, Mary and Stanislaus, with the support I failed to give them. They are victims too. And they do not deserve that. And, please, also pray for me." He sat down.

On leaving the church after unvesting, Frederick Patrick observed the Krakowski parents. They were surrounded by fellow parishioners.

Father Curley increased the car's speed as they left the outskirts of Posnan. The bishop sat in back, his grief transparent.

CHAPTER THIRTY-SIX

The day itself was funereal. Low dark clouds and cold drizzle delayed the Thursday dawn and then held it through the morning. The bells of St. Boniface, their mournful tolling magnified by the dense atmosphere, vibrated the small town of Ilanz. Their sound labored down through the town and out over the mighty river, and it pulled up the high cliffs above the town. The line of cars, lights on, left the funeral home and drove slowly along the river road. The lead car turned upward into the hill and led the way to the church. The cars filed into the parking lot while the hearse stopped at the main entrance.

The family of Charles Mueller entered the church and straggled to the front right pews. Karl and Eva Mueller, Charles's parents, sat at the center aisle of the front pew. Their children and grandchildren sided and backed them. A few cried, but most were already drained. On the opposite side of the aisle, clergy filled the front third of the church. Dozens of flower arrangements spread their odor throughout the church.

The celebrant, Father Max Bauer, like Charles, a priest born in the parish and a family friend, waited at the church entrance with the deacon and subdeacon, classmates of Charles. They watched through the open door as the pall bearers lifted the casket, carried it up the entrance steps and placed it on its rolling platform. The funeral director wiped it dry.

After reciting the introductory prayers, the celebrant circled the casket, first with billowing incense, then with sprinkles of holy water. He and his retinue followed the casket to the front while the incense fought and conquered the flowers' fragrance. Frederick Patrick Sweeney waited in the sanctuary. He stood, fully vested and attended, at the sedilia.

Tony Corsalini read the first scripture reading from the book of the prophet Amos. His voice broke as he read:

Yahweh is his name....

He is trouble for those who turn justice into wormwood, throwing integrity to the ground; trouble for those who hate the man dispensing justice at the city gate and detest those who speak with honesty.

Sister Margaret McDonough was dressed in a black suit. The silver medallion of the Franciscan Order glistened against her white blouse. Her face was ashen as she read the second reading from scripture, St. Paul's lyric celebration of immortality and victory over death:

...all of us are to be changed, in an instant, in the twinkling of an eye, at the sound of the last trumpet. The trumpet will sound and the dead will be raised incorruptible, and we shall be changed...

Tears choked a halt. She recovered and continued.

When the corruptible frame takes on incorruptibility and the mortal immortality, then will the saying of scripture be fulfilled: "Death is swallowed up in victory. O death, where is your victory? O death, where is your sting?"

Bishop Sweeney looked out at the congregataion. His eyes swept the faces of the assembled priests. The face of John Crocker checked his course. I wonder what you are thinking, he thought. I wonder if you comprehend that this man died for your sin? Frederick Patrick could read no remorse or guilt on Crocker's impassive face. What strange

248

configuration of conscience and emotion can leave you with no awareness?

The deacon read the gospel, from the apostle John.

Jesus answered them. "The hour has come for the Son of Man be be glorified. I solemnly assure you, unless the grain of wheat falls to the earth and dies, it remains just a grain of wheat. But if it dies, it yields a rich harvest.

When the reading had finished and the deacon was seated, the bishop moved to the sanctuary center to deliver the homily.

"My dear friends in Christ," he began. He acknowledged and expressed sympathy to the Mueller family, and then continued, his words coming slowly. "Father Charles Mueller made me feel uneasy whenever we met. He was my uninvited critic." The bishop paused. "Because of him I will henceforth treasure all critics. Yet, until his death, I did not treasure Father Mueller." Frederick Patrick felt the beads of perspiration as they formed on his forehead.

"Father Mueller was a prophet, a man of courage, like the prophet Amos in the first reading from scripture, who spoke the truth as he saw it regardless of consequences. You are looking at the consequence of his courage, his casket." The bishop stopped for breath.

"There are two parts to this tale of courage and death. The first part is the story of a young boy who was abused by a priest. That priest was not Father Mueller. When I learned of that abuse, I was afraid for the institution of the Church. I protected the institution, but failed to care for the boy. That boy, troubled to the point of madness, is the boy who shot Father Mueller." His glance caught Crocker's face, still impassive.

"The other part of this story is Father Mueller's. He pointed out to me, without anger, without self-interest, other occasions where I had placed the institution ahead of individuals. I know now that I began to change when he challenged me, a change that began with the uneasiness he imposed on me. He made me question what I had never questioned. At the time, I was unable to face the sensible answers to those questions, and buried them in anger. In my anger I exiled Father Mueller from Mill Valley. Unknowingly, I sent him to face a gun whose hammer I had helped cock. But, the gun that killed Father Mueller also extinguished my anger, and enabled me to face the answers that were alive, though buried in me."

Sweat was now a river on the bishop's face. He was propelled onward by the force of his newfound insights.

"A kind and learned cardinal of the Church, Cardinal Alberto della Tevere, once told me that our Church institution does not always reveal the hand and face of Christ. I did not understand him then. I do now. He said what Father Mueller said. The institution does not have rights. Only people have rights. The Church is the people, not the institution. A bishop should respect those rights above all else if he is to be true to Christ." The bishop paused again. "I will remember Father Mueller until the day I die as a great benefactor to me. From the dawn of time, it seems, we have killed our prophets, men selected by God to speak for Him. But a prophet's death is never in vain. And I dedicate my remaining years to the lesson this prophet has taught me. Charles Mueller is the grain of wheat who has fallen to the earth and died. I am proud to consider myself but one of the newly-formed ears of wheat in the abundant harvest that will rise from his death. As you pray for him, please also pray for me." Frederick Patrick sat down. Exhausted and drenched with sweat, he had talked less than five minutes.

250

It happened all of a sudden. Josephine was once again the bishop's secretary. She fretted. Bishop Sweeney seemed so preoccupied. He had failed to compliment her on her work. And he had even written several letters longhand. Why would he do that? Didn't he trust her? And why would he fly to San Antonio in the morning and return the same day? She had put the call into the airline to make the reservations.

Frederick Patrick began his letter: "Eminenza," stopped, tore the paper in bits, and began again. "Dear Alberto." New emotions surged from freshly-tapped wells as he recounted the recent events. Perhaps, he thought, my need to confess has not been satisfied. Or is it that I need to hear someone say it's all right before I move on? He finished the letter and placed it in an addressed envelope.

Josephine appeared in the doorway. "Detective Pritzlaff, from the John's Point Sheriff's Department is here for his appointment, Bishop."

"Thank you, Josephine. Would you mail this letter by the fastest means possible, please. And show the officer in." He handed the letter to Josephine.

Detective Pritzlaff was quietly courteous, even apologetic for his intrusion. There were some answers yet needed to complete his investigation. While there was perhaps no need, should it make the bishop more comfortable, he would be welcome to have an attorney present. The bishop declined the offer.

"The parents of the young man who shot Father Mueller told me that he had been abused by a priest. I understand that

251

you acknowledged that fact in your sermon at the boy's funeral?"

"It's true."

"How do you know that?"

"The priest admitted it to me when I confronted him."

"Did you report it to the police?

"No, I did not."

"Why not, Bishop? Sexual assault is a felony. Surely you know that?" His voice remained quiet and kindly.

"Yes, I knew that. But I also knew that he would deny it to the police."

"Father Crocker would deny it?"

"Yes."

"How do you know that?"

"He told me so. And to my mind at the time that rendered anything I might tell the police hearsay. I thought that the complaint could only be filed by the boy."

"That's not entirely correct, Bishop. What's more, it's easy to understand why a boy would bury abuse of this kind rather than reveal it. His parents might have helped him there." He paused.

"You're right. And that's where I was certainly wrong. It never occurred to me to counsel them in that direction when they came to me. I was so concerned about public scandal that I made light of the boy's needs. I did what I could to impose silence on those good people. I should have encouraged them, assisted them, even directed them to the police. I would do that if I could do it over."

"So, you do understand what you should have done, Bishop." The detective's quiet voice was as penetrating as his eyes. "Your silence and that of the parents abetted the perpetrator of a serious crime. That is never lawful, even if the perpetrator is a priest." He stopped for a moment. "There is nothing we can do now. The boy is dead. Prosecution wouldn't stand up in court. The boy's parents have chosen not to pursue the matter further. You are fortunate there. And I have sufficient data to complete my investigation." He rose. "Thank you again for your time."

The bishop stood. They shook hands. The deputy, more accustomed to the disquiet both men felt from the lack of closure, left quietly. The bishop sat down again. How much more is there to life that our pithy syllogisms don't teach? he asked himself.

Acceptance and approval came to the bishop ten days later. Josephine brought the morning mail, sat, primped, and waited for the bishop to dictate replies. The bishop read the opened letters in their stacked order, replied, deferred, delegated or tossed as he elected. Midway through the stack his eyes caught the return address on the next envelope. "Josephine, I'll need some time for the next letter. You can type the ones we've finished. I'll call you when I'm ready."

He was anxious. The time it took Josephine to close her notebook, rise and leave seemed unusually long. He had the letter out of the envelope before the door closed.

Caro Federico,

My heart is heavy with your pain. And I feel terribly the loss of your Father Mueller. Still, I cannot put in words the joy your letter has given me. My mind and my heart both tell me that you now have the full spirit of Il Cero. You must not distress yourself over what is

253

past and cannot be altered. Thank the Lord, rather, for his kindness in taking the scales from your eyes as he did for the blind man. I have prayed for this.

We will talk more of it when I see you. I apologize for this short notice. It has taken our members longer than we anticipated to prepare for our meeting. Il Cero will meet at a convent in Fiesole on October 21st. My sister is superior there and we will have total privacy. The convent has a beautiful setting overlooking Firenze. We expect to be there for about ten days.

I would be pleased if you will fly to Rome on October 17 or 18 and be my guest. Please send me your schedule and I will meet your plane. We can talk, it is important I think, and then drive together to the meeting.

Again, I rejoice for you. Until I see you, my prayers for a safe journey.

Affectionately,

Alberto

CHAPTER THIRTY-EIGHT

"Sister Margaret will see you now, Bishop. Come this way, please." The elderly nun, a receptionist in her retirement, led Frederick Patrick down a varnished and polished hall toward an open door. At the sound of their steps Maggie had risen and was half across the office when the bishop was ushered in.

"Good afternoon, Bishop." She did not offer to shake his hand. "Please have a seat here." Maggie pointed to a chair in a seating arrangement reminiscent of her office at Padua. "What can I do for you?"

"Thank you for seeing me, Sister Margaret. I'm here to give you an apology." He spoke slowly and softly, deliberate in his intent to say all that needed saying.

Maggie remained silent, placing the burden of her silence on the bishop.

"I can see now that I did you an injustice," Frederick Patrick continued, accepting the burden. "It was I who gave impetus to the trustees who terminated your employment at Padua. At the time I thought my action best served the Church. I was wrong."

Maggie's silence became a scream that continued to dominate the room.

"Your feminist convictions did not seem to accord with those of the Church leadership. I know now that I was anxious for the institution of the Church. I still do not know if you are correct in your views, but if any institution should protect your right to say them it is the Church. I now realize that to fear dissent, if one claims the truth, is cowardly. To suppress dissent without due process is both cowardly and unjust. I would prefer to disown my status as bishop if either of those were the only alternative." The bishop stopped and waited.

Maggie was the first to break the now equalized burden of silence. "It was kind of you to come, Bishop. For what it's worth, I accept your apology. It is not unlike the words you spoke at Father Mueller's funeral. And, I've heard from Tony and Kate Corsalini that you have spoken to them in a similar vein in San Antonio. Where do we go from here?" The words spoke her conscience, but not her feelings. She remained coldly distant.

"If I can achieve your reinstatement at Padua without working similar injustice in the process, I will do my utmost, if that's your wish." It was a question.

Maggie's response was immediate. "I think not, Bishop. The Corsalinis and Father Mueller were my dearest friends in Mill Valley. I'm not ready to go back and face the cold void of their absence. Besides, the people now in the administration at Padua have accepted their positions in good faith. Even should Monsignor Deutschman give up his position voluntarily, the other newly-appointed administrators are unacceptable to me. I don't need the hassle that dealing with them would bring into my life just now." The memory of John Winslow Alexander traced its effect on her face.

"Is there any other way I can make recompense?"

Maggie now felt the burden. She struggled for the unqualified forgiveness dictated by her faith. She could not reach it. "Can you bring Father Mueller back?"

The bishop moved his hands, a gesture of helplessness. "I'm sorry, I won't take up any more of your time, Sister Margaret. It was good of you to see me." He rose. "I almost forgot," he said, reaching for his briefcase. He opened the briefcase and took out a sheaf of envelopes. "These were in Father Mueller's desk. They are sealed, stamped and addressed to you. Why he did not mail them, I do not know. I have chosen to be his postman." He handed the sheaf to Maggie.

Maggie reached for the envelopes and cradled them gently in slender fingers. "Thank you," she said softly. She stood and walked him to the door. "Goodbye, Bishop." She extended her hand.

The bishop grasped it in a silent handshake. He had kept the taxi waiting. As they drove back through the city, the driver's chatter seemed warm, and welcome.

Frederick Patrick awoke as dawn touched the nose of the airplane somewhere over the Atlantic. His thoughts turned to Il Cero.

CHAPTER THIRTY-NINE

Memorial Day

Angela and Angelo Pannetto had been named by a romantic angel-stricken mother. Angelo, a bachelor, lived on the home farm. On Charles's death he had asked Angela to come home. She came, much to the disappointment of several housekeeper-seeking priests.

"I'm just not ready," she told them. "I'm gonna smell my own roses and plant my own garden."

This morning, Angela carefully backed her car up to the back porch of the homestead. The porch had been enclosed with glass years earlier by Angelo. It had been home to his garden seedlings, a place to give them a jump-start on summer. The porch was now in Angela's domain. She propped the back door open with a stick and carried two cardboard crates of geranium plants, and one of alyssum to the car. Then she laid spade, hoe and trowel next to the plants.

Her short, bushy-haired brother came onto the porch. "You need some help, Angela?"

"I got everything loaded already, Angelo," she responded. "You take care now. I'll see you tomorrow. Your dinner is in the refrigerator." She walked over and hugged him.

She drove the car down the country road and pulled onto the highway heading north along the great river. The brown waters sparked fire from the bright sun. A few clouds dropped blotches of shadow on both the land and the water. Twenty miles up the road Angela took a paved township road and minutes later pulled into the yard of Eva and Karl Mueller.

Eva and Karl appeared as Angela arrived at the kitchen door.

"Do I know this beautiful lady?" Karl charmed. "Introduce me, Eva."

"Introduce yourself, you old lecher," Eva said, laughing.

Karl transferred the plants from Angela's trunk to the back of his van while the women chatted.

"I've got tools, Angela," he said. "We can leave yours in your trunk."

"That's fine, Karl. But take my trowel, would you please. I got so used to it."

Karl opened the van's doors for the women, then drove north up the river highway to Ilanz. They stopped at a small home on the outskirts. Gertrude was watching for them. Gertrude had come home to Ilanz after Frosty's funeral. Her brothers and sisters were all on local farms and each had invited her to live with them. She had elected to live alone, near the church. Karl opened a door for her.

Karl took a road east into the bluffs. The road hairpinned up the steep slopes of a valley. At the summit Karl took a lane that angled back to the bluff's edge. There lay St. Boniface Cemetery, its acres bordered on the front and sides by a black wrought iron fence, and on the back by the sweeping view of the great river's wide valley.

Karl opened the wrought iron gate and drove to the grave site. A newly-hewn red granite monument stood at the head of the grave's bare soil. The lettering, dictated by Charles in his will, read simply,

Charles Mueller
Priest
1940-1984

Eva, Angela and Gertrude took hoes and trowels and began preparing the ground. Karl carried rolls of sod and the flower boxes from the van. While the women worked on the grave, Karl took a weed whacker to the high grass of the fence lines. A pheasant whirred up, startling them.

Eva and Gertrude rolled the sod out onto the grave while Angela worked the soil in front of the plot. A heavily-laden cumulus dropped sprinkles on them.

"Oh," Gertrude remarked, "Father Frosty would say that God is blessing us with his own holy water."

"Doesn't that sound like Father Frosty though," Eva responded. Eva and Gertrude handed geraniums and alyssum to Angela.

"You brought so many plants, Angela," Eva commented.

"I suppose," Angela replied. "I just wanted to be sure we have enough for Father Frosty's grave too.

"Look at the eagles," Eva said suddenly. "Charles so loved the eagles."

The others raised heads to look at the pair of eagles soaring above them.

Karl had returned, put the weed whacker into the van, and stood watching. "We'll drive right over to Father Frosty's grave at Burnside when you finish here," he said. "We don't want those beautiful plants to wilt."

"I wonder what it's really like up there," Angela said after planting the last geranium. She pointed to the sky.

261

"Well, Father Frosty's got a garden. You can bet on that," Gertrude answered.

"And a lake, Gertrude, don't forget that," Karl added. "And you can be sure God keeps it frozen for him 365 days of the year."

"And Father Charles?" Angela asked softly.

Karl was thoughtful for a few seconds. "Oh, he's probably measuring himself against a big cliff, or fishing with Father Frosty, or walking a fence line, or, right now, soaring up there with those eagles."

Eva's eyes teared.

Karl knelt on one knee in front of the grave and used his index finger to punch a shallow hole in the loose soil between the flowers. He reached into his jeans and withdrew a palm full of wheat. He selected a single grain, dropped it into the hole and smoothed soil on top. He looked up at Eva's tears. His own eyes filled.

Dear Reader,

If you enjoyed this book please recommend it to your friends. Thanks. Steve.

Order Form

Please send _____copies of *Unless A Grain Of Wheat* to:

Name: _____

Address: _____

City: _____State: ____Zip: _____

Phone:(_____)_____

Price:
1-5 copies: $14.95 ea.   # copies____ $ _____
5-10 copies: $13.50 ea.   # copies____ $ _____
10 plus copies: $12.00 ea.   # copies____ $ _____

Postage and Handling:
$3.00 first book; $1.00 each additional $ _____
                                      Total: $ _____

Please make check payable to Wind-borne Publications.
Send order and check to:

Wind-borne Publications
P.O. Box 733
Hales Corners, WI 53130

Dear Reader,

If you enjoyed this book please recommend it to your friends. Thanks. Steve.

Order Form

Please send _____copies of *Unless A Grain Of Wheat* to:

Name: _____

Address: _____

City: _____State: ____Zip: _____

Phone:(_____)_____

Price:
1-5 copies: $14.95 ea.   # copies____ $ _____
5-10 copies: $13.50 ea.   # copies____ $ _____
10 plus copies: $12.00 ea.   # copies____ $ _____

Postage and Handling:
$3.00 first book; $1.00 each additional $ _____
                                      Total: $ _____

Please make check payable to Wind-borne Publications.
Send order and check to:

Wind-borne Publications
P.O. Box 733
Hales Corners, WI 53130